Handbook for Plastics Processors

Handbook for Plastics Processors

J. A. Brydson

Published in association with
The Plastics and Rubber Institute

Heinemann Newnes

Heinemann Newnes
An imprint of Heinemann Professional Publishing Ltd
Halley Court, Jordan Hill, Oxford OX2 8EJ

OXFORD LONDON MELBOURNE AUCKLAND SINGAPORE
IBADAN NAIROBI GABORONE KINGSTON

First published 1990

British Library Cataloguing in Publication Data
Brydson, J. A. (John Andrew), *1932–*
  Handbook for plastics processors.
  1. Plastics. Processing
  I. Title
  668.4

ISBN 0 434 90200 4

Printed and bound in Great Britain
by Butler & Tanner Ltd, London & Frome

CONTENTS

PREFACE

Successful manufacture of an article from plastics, once its purpose and its requirements in service are established, requires knowledge of

- the plastics materials from which it might be made, their nature and the ways in which processing may affect their properties;
- the processing methods available for its manufacture;
- how to evaluate its properties and to apply effective quality control.

The first object of this handbook is to meet these needs of the processor with a concise summary of essential background knowledge. Its second object is to provide a supporting body of information on specific practical techniques, procedures and problems. Thus the materials survey is followed by a simple identification scheme which can easily be adapted for specific use on the shop floor; the section on processing includes a chapter on trouble shooting with the major processes and a brief discussion of design principles and problems.

The general data section covers such topics as systems of measurement, thermal properties of polymers, specific gravities, oxygen index, heat stabilities, flow properties, shrinkage data and hydraulic and electronic symbols. What is dis-

tinctive in this handbook is that the tables of data, which I believe to be those most likely to be required by the processor, are normally preceded by an explanation of the meaning of the data and how it may be used. All too often data (including data in the form of a computer program) is used without a true appreciation of its scope and limitations; this can introduce serious errors, which I hope these annotations will help to avoid.

Finally, in the appendices are some specimen calculations of interest to processors, sources of further information including references to specialized works on specific topics, and some general comments on health and safety aspects of polymer processing.

Clearly I have had to be very selective in material in order to keep to a handbook format. My general criterion has been whether I consider a topic or item useful to the plastics processor either in broadening his general knowledge and understanding of the subject or in providing frequently required data in a compact and easily retrievable form under a single cover.

I hope that the resulting handbook will prove to be one that the user will want to keep within easy reach.

Brent Eleigh, Suffolk   John Brydson

ACKNOWLEDGEMENTS

While one must of course acknowledge the work of the thousands of scientists and technologists who first acquired the data and knowledge upon which this book is based I should like to pay particular thanks to the Plastics and Rubber Institute and to the Plastics Processing Industry Training Board for their permission to make wide use of diagrams from their publications. While mentioning diagrams I have to confess to problems in acknowledging the true source of many of them. Many depict basic processing features which I have found reproduced, in identical or similar form, in many publications including trade literature so that one cannot be sure of the original source. In reality it must be the, usually anonymous, developer of the particular process. I should also like to give personal thanks to Susan Davey and Matthew Hepburn of the library of the Polytechnic of North London for once again giving generous help in searching out information and data.

MATERIALS

1

THE NATURE OF PLASTICS AND COMPOSITES

INTRODUCTION

The types of plastics material that are available commercially may be counted in hundreds, and within each type there are many grades. Most of these materials are man-made (although there are a few natural plastics) and most belong to a class of chemicals known as polymers. The chemist has discovered ways of tailoring the chemical structure of these polymers in order to modify their properties. At the same time the plastics technologist has also achieved a measure of control over these properties by the use of additives such as fillers, antioxidants and plasticizers, and by altering the process conditions. Since, in a broad sense, the use of additives and control of processing conditions affects the chemical nature and hence the properties of plastics materials it is useful even for the non-chemist to have some understanding of the materials which he or she uses.

This chapter assumes no knowledge of chemistry. However the reader already familiar with the concepts of atoms, elements and molecules may go straight to page 5.

ATOMS, ELEMENTS AND MOLECULES

Plastics, like everything else that exists materially on earth, consist of chemicals. In order to understand their properties it is first necessary to understand something of the way in which chemicals are made up.

If we take a glass of water and drink exactly half of the water, that remaining in the glass will still be water. If we drink half of the water that is left, once again the remainder will still be water. This process may be repeated in theory many times although in practice it will become more and more difficult to drink exactly half of the remaining water. Eventually a point is reached when, if we try to divide the contents, the fractions will no longer be water. When this stage is reached we will have one *molecule* of water. If it is further divided we find that the water molecule is made up of two *atoms* of hydrogen and one atom of oxygen.

Hydrogen and oxygen are examples of *elements* of which about ninety occur naturally with some additional ones having been made in the laboratory. All of the chemicals with which this earth is made are, in turn, made up from these elements. Other well-known examples of elements are nitrogen, iron, lead, copper, carbon, silicon and uranium.

Atoms are extremely small, the smallest, hydrogen, weighing only about 1.7×10^{-24} grammes. The ratio of a spoonful of sugar to the size of the earth is about the same as that of a hydrogen atom to that of the spoonful of sugar.

In chemistry it is often important to consider the size of atoms and molecules but it is often unwieldy to quote in terms of grammes. For this reason it is more convenient to talk in terms of relative masses. If hydrogen is given a *relative atomic mass* of 1 then that of oxygen is about 16, of carbon 12 and of nitrogen 14. (N.B. Although originally the relative atomic scale was devised based relative to hydrogen, current practice gives a value of 12 to a particular form or isotope of the carbon atom and other atoms are given atomic masses relative to this.) In a similar way it is possible to give an estimate of the *relative molecular mass*. For example the relative molecular mass of water will be the sum of the relative atomic masses present. With two atoms of hydrogen (relative atomic mass of 1) and one of oxygen (16) the relative molecular mass of water will be 18. Similarly the relative

molecular mass of carbon dioxide (one carbon and two oxygen atoms) will be $12 + 2 \times 16 = 44$. Until recently relative atomic mass and relative molecular mass were known respectively as *atomic weight* and *molecular weight*. These latter terms are still widely used in industry.

Because of the complexity of real chemicals it is also convenient to develop a form of shorthand when writing the formulae for them. This is based on the allocation of a letter(s) as a symbol for each element. Some common symbols are:

Carbon	C
Chlorine	Cl
Helium	He
Hydrogen	H
Nitrogen	N
Oxygen	O
Sulphur	S
Silicon	Si

(A complete list of the elements together with their symbols and relative atomic masses is given in the table on page 160.)

Using this shorthand we may write the *empirical formulae* of water as H_2O, of carbon dioxide as CO_2, of methane as CH_4 and silicon tetrachloride as $SiCl_4$.

We could also write the empirical formula of, for example, ethyl alcohol (the 'active' constituent of alcoholic drinks) as C_2H_6O but this gives little information as to how the molecule is put together. To a chemist it is more useful to indicate chemical groups that are present by writing the formula as C_2H_5OH. The C_2H_5 is known as an ethyl group (which has certain characteristic properties) and the OH as a hydroxyl group which also has certain characteristics.

In order to understand chemical formulae, including those of plastics materials, it is also useful to understand something known as *valency*. It has been established that there are a definite number of links which an atom can make to other atoms. In the case of hydrogen it is 1, with oxygen 2, with carbon 4 and with chlorine 1. The valencies of these atoms are 1, 2, 4 and 1 respectively. Some atoms can take more than one valency such as nitrogen which may take valencies of 3 or 5.

It will be seen that the following structures for water and methane satisfy these valency rules.

$$H-O-H \qquad \begin{array}{c} H \\ | \\ H-C-H \\ | \\ H \end{array}$$

(water)   (methane)

The links between atoms are referred to as *bonds*. In the two examples given above, all of the bonds are known as *single bonds*. With some molecules it is necessary to have *double bonds* to satisfy the valency rules. For example, the gas ethylene, from which polyethylene is made has the following structure

$$\begin{array}{cc} H & H \\ | & | \\ C & = C \\ | & | \\ H & H \end{array}$$

Molecules also exist with *triple bonds* (such as acetylene).

$$CH \equiv CH$$

A final complication to be dealt with here is the fact that bonds between atoms in a molecule have definite *bond angles* to each other. Thus, although the formula of methane may be written as above it would be wrong to believe that all of the atoms were really in one plane with the bonds at right angles to each other. It is more accurate to picture the carbon atom in the centre of a tetrahedron with a hydrogen atom at each apex, with bond angles of about 109°. The non-planar nature of many molecules is of particular significance in plastics manufacture.

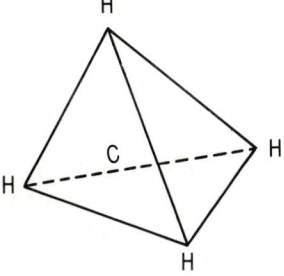

PLASTICS AND POLYMERS

The base materials from which plastics are made consist of large molecules. Most have a relative molecular mass or molecular weight of at least 100 000, and in the case of some acrylic sheeting over 1 000 000. Although some plastics are based on natural materials such as cellulose most are made synthetically from fairly simple small molecules.

In many cases only one type of small molecule may be used to make the material, sometimes two and very occasionally more than two. The simplest example is ethylene which has the structure

$$\begin{array}{cc} H & H \\ | & | \\ C & = C \\ | & | \\ H & H \end{array} \qquad \text{which may also be written } CH_2 = CH_2$$

It will be noticed that this molecule possesses a carbon–carbon double bond. Such a bond may be partially opened (by materials known as initiators such as peroxides) and the individual ethylene molecules join together to form a long chain molecule of polyethylene of which a short section would have the structure.

$$-CH_2-CH_2-CH_2-CH_2-$$

The process of joining together such simple molecules is known as *polymerization*. Ethylene is an example of what is known as a *monomer* and polyethylene an example of a *polymer*.

Many polymers used in plastics are made by a similar process and some of the most important are given in Table 1.1. All of the examples of monomers given in Table 1.1, except formaldehyde, have structures related to ethylene, but in which one or more of the hydrogen atoms in ethylene has been replaced by other atoms or groups of atoms. In the case of the polymerization of formaldehyde the process involves the opening of a carbon–oxygen double bond. Although these materials are amongst the most important plastics (the first four alone account for over three-quarters of the market) polymers may be made in other ways.

Table 1.1 *Some important polymers and their monomers (all polymerizations involving opening of double bonds)*

Monomer	Polymer
$CH_2 = CH_2$ Ethylene	$- CH_2 - CH_2 -$ Polyethylene
$CH_2 = CH$ | CH_3 Propylene	$- CH_2 - CH -$ | CH_3 Polypropylene
$CH_2 = CH$ | Cl Vinyl chloride	$-CH_2 - CH -$ | Cl Polyvinyl chloride
$CH_2 = CH$ | $\emptyset$ * Styrene	$- CH_2 - CH -$ | $\emptyset$ * Polystyrene
$CF_2 = CF_2$ Tetrafluoroethylene	$- CF_2 - CF_2 -$ Polytetrafluoroethylene
CH_3 | $CH_2 = C$ | $COOCH_3$ Methyl methacrylate	CH_3 | $- CH_2 - C -$ | $COOCH_3$ Polymethyl methacrylate
$CH_2 = O$ Formaldehyde	$- CH_2 - O -$ Polyformaldehyde (Polyacetal)

* The symbol $\emptyset$ represents a ring of structure

$$-C \overset{CH-CH}{\underset{CH = CH}{\diagdown}} CH$$

As with the other polymers mentioned in Table 1.1 polyformaldehyde is made by *double bond polymerization*. This is characterized by the opening of double bonds and joining together of the monomers without elimination of any small molecule as a side product.

Polymers may also be prepared by opening up and linking of small ring-shaped molecules (*ring-opening polymerization*). Examples are given in Table 1.2.

Table 1.2 *Polymers produced by ring-opening polymerization*

Monomer	Polymer
$\overset{O}{\overset{\diagup\diagdown}{CH_2-CH_2}}$ Ethylene oxide	$- CH_2 - CH_2 - O -$ Polyethylene oxide
$(CH_2)_5 \overset{CO}{\underset{NH}{\big|}}$ ϵ–Caprolactam	$- (CH_2)_5 \, CO.NH -$ Nylon 6
$(CH_2)_5 \overset{CO}{\underset{O}{\big|}}$ ϵ–Caprolactone	$- (CH_2)_5 \, CO.O -$ Polycaprolactone

As with double bond polymerization no small molecules are eliminated.

Ring-opening polymerizations are sometimes considered as special examples of *rearrangement polymerizations*, processes in which the molecules involved become arranged but without any elimination. A well known, but somewhat complicated example, of a rearrangement polymerization is the reaction of a di-isocyanate and a di-alcohol (diol) to form a polyurethane.

$$OCN(CH_2)_6NCO + HO \cdot (CH_2)_4 \cdot OH \rightarrow \; \sim OCNH \cdot (CH_2)_6NHCOO \cdot (CH_2)_4O \sim$$

Many polymers are made by *condensation polymerization*. An important example is the reaction of adipic acid and hexamethylene diamine to give the polyamide known as nylon 66. In this case the acid ($-COOH$) groups react with the amine ($-NH_2$) groups to form an *amide* link with the elimination of water.

$$H_2N \, (CH_2)_6NH_2 + HOOC(CH_2)_4COOH$$
$$\rightarrow \; \sim NH \cdot (CH_2)_6NH \cdot OC(CH_2)_4CO \cdot \sim + H_2O$$

Polyesters such as polyethylene terephthalate are also made by condensation polymerization. All of the polymers depicted in Table 1.2 may also be made by condensation polymerization of suitable starting materials and even a structure the same as polyethylene given earlier in the section can be made by condensation polymerization of diazomethane. The small molecules used in rearrangement and condensation polymerizations are usually known as *intermediates* rather than monomers.

To recapitulate, synthetic plastics are prepared by the process of polymerization of which there are the following main variations:

1 double bond polymerization;
2 ring-opening polymerization;
3 rearrangement polymerization;
4 condensation polymerization.

Finally, in this section, it should be noted that while plastics are based on polymers, polymers are not necessarily plastics. Some polyester polymers are plastics, some are rubbers, some are fibres, some surface coatings and others adhesives. Sometimes one particular polymer, such as nylon 66, may be both a fibre or a plastics material according to the way in which it is processed.

HOMOPOLYMERS AND COPOLYMERS

When one monomer is polymerized alone the product is sometimes referred to as a *homopolymer*. Thus polyethylene and all of the other polymers listed in Table 1.1 may be considered as homopolymers.

Useful products may, however, often be prepared by feeding two or more different monomers into the polymerization reactor so that an individual polymer chain will contain residues of each of the monomers. The process is known as *copolymerization* and the products as *copolymers*. (This is distinct from mixing two homopolymers where each individual chain will contain only the residues of one monomer species.)

Where just two monomers have been used the term *binary copolymerization* may be used, although in practice the products are often simply spoken of as copolymers. Where three monomers have been used the process is known as *ternary copolymerization*, with the products commonly referred to as *terpolymers*.

In a copolymer the arrangement of the monomer units can take many forms. Many important commercial materials approximate to a *random copolymer* arrangement with the monomer units randomly distributed along the chain (and a statistically random between-chain distribution is also usually inferred). For a (binary) copolymer made up from two monomers A and B the arrangement might be of such a form as

AAABBABBABBABABBBAABAAABBABBBBAAABABAABBABBBAABA

Another possibility, although rarely encountered in practice, is the *alternating copolymer* of form

ABABABABABABABABABABABABABABABABABABABABABABABAB

In a *block copolymer* the units are separated. The structure

——AAAAAAAAAAAAAAAAAAAABBBBBBBBBBBBBBBBBBBB——

is a example of a *diblock copolymer* and

——BBBBBBBBBBBBBBBBAAAAAA——AAAAAAAAAAABBBBBBBBBBBB——

is known as a *triblock copolymer*.

Block copolymers are particularly important with thermoplastic rubbers. Intermediate between a random and a block copolymer is the *tapered copolymer*. In these polymers one end of the chain is rich in the units of one monomer but less rich at the other.

AAAAABAAAABAABAAAABBAABAABBAAABABAABBABABBABBBABBB——
——BABBBABBBBBABBBBBBBBBB

Another form of copolymer is the *graft copolymer*. In this case a side chain, different in nature to the main chain is grafted onto the main chain. Both the side chain and the main chain may themselves be made up of one or more monomer units.

THERMOPLASTICS AND THERMOSETTING PLASTICS

All of the examples of plastics materials given on page 6 consisted of long chain molecules known as *linear polymers*. At a sufficiently low temperature these polymer molecules will be immobile and in the mass the polymer will be a solid. On heating the material, however, the molecules will absorb energy and eventually a point will be reached where they are able to rotate about their bonds and, if it were possible to see them, would appear to be like a mass of wriggling worms. In effect, this mass will be molten and on application of a stress to the material it will flow and may be shaped. If the material is then cooled down it will harden. Materials that soften on heating and harden on cooling are known as *thermoplastics*. Their behaviour may be expressed schematically by a diagram of form

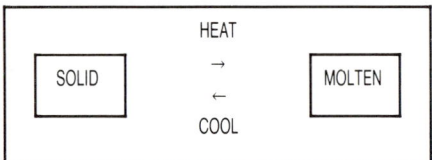

Although in theory the heating–cooling process may be repeated indefinitely, real materials do tend to degrade on heating and the number of times that the process may be repeated is limited.

Until the 1950s thermoplastics were in the minority amongst plastics. Far more important were the *thermosetting plastics*. These are best understood if we consider the reaction between glycerol and an acid of general structure HOOCXCOOH (X can simply represent the rest of the molecule which does not take part in the reaction to be described below). Glycerol has three hydroxyl groups, and so reaction with acid groups can take place in three directions, and eventually a three-dimensional network structure is built up.

$$CH_2-OH \quad HOOCXCOOH \quad HO-CH_2$$

~COOH HO–CH CH–OH HOOCX~

$$CH_2-OH \quad HOOCXCOOH \quad CH_2$$

HO OH

CH$_2$ HOOCX~

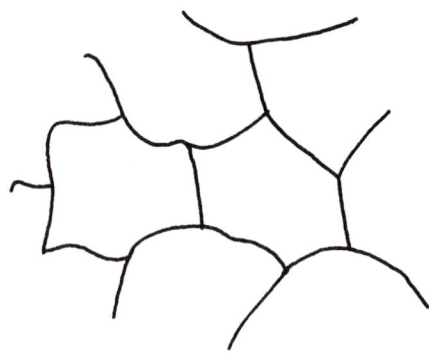

Such a system will not melt on heating. Since the first of the commercial polymers of this type also required heat for the three-dimensional (cross-linking) reaction to proceed, such plastics became known as thermosetting plastics. Although some cross-linked polymers may now be prepared by reacting at room temperature, the term thermosetting plastics has been retained to describe the type of material rather than the method of polymerization.

Important thermosetting materials include the phenolic resins, aminoplastics such as the urea-formaldehyde and melamine-formaldehyde plastics, unsaturated polyesters used for reinforced plastics and the epoxide resins.

AMORPHOUS THERMOPLASTICS AND THE GLASS TRANSITION TEMPERATURE (T_g)

Even when no additives are present polyethylene, polypropylene, the polyacetals and nylon 66 are translucent to opaque whereas unfilled polystyrene, PVC and polymethyl methacrylate are transparent. The reason for the difference is that the first three polymers are crystalline and the second three amorphous or glassy. Whether a polymer is amorphous or glassy has an important influence, not only on the properties of the end product, but also on processing behaviour.

In an amorphous polymer the molecules do not arrange themselves in any real order. In the molten state, and not subjected to external stresses, they tend to coil up in a statistically random fashion. In such a molten state the molecules have sufficient thermal energy that the molecular chains move by rotating about their bonds rather like wriggling worms. If such a mass is slowly cooled a temperature is reached at which the molecules do not have the energy to move in this way and the mass hardens. This temperature is known as the *glass transition temperature* and is universally given the abbreviation, T_g. (Note: Strictly speaking, hardening occurs over a range of temperatures and the value obtained for T_g will depend on the test method used.)

With all but the lowest molecular weight polymers there is not a sudden transition between solid and liquid, but the material goes through a rubbery range on heating and cooling. In this rubbery range, molecules coil and uncoil but do not slide past each other on application of an external stress because of entanglements that are present. As the molecular weight increases the level of entanglement may reach the stage that the material starts to decompose before the polymer melts. This occurs, for example, with high molecular weight acrylic sheeting such as Perspex while lower molecular weight polymers of the same structure may be processed without undue difficulty.

A typical temperature–molecular weight diagram for an amorphous or glassy polymer is given in Figure 1.1(a).

CRYSTALLINE THERMOPLASTICS AND THE CRYSTALLINE MELTING POINT (T_m)

In order to understand the properties of crystalline polymers, it is first necessary to say a few words about the nature of crystallinity in polymers, which is somewhat different than in simple crystalline solids. With non-polymeric crystals such as those of common salt, Epsom salts and copper sulphate the crystals assume definite forms bounded by perfectly plane (flat) faces which have characteristic and constant angles to each other. The simplest of these is the cubic system displayed, for example, by common salt, $NaCl$. In such crystalline structures the atoms or molecules are arranged in a definite way with very few imperfections. Such a level of perfection in molecule arrangement is very difficult with high molecular weight polymers because of the difficulty of the large molecules being able to take up a highly regular arrangement for the whole of the molecule. What does however occur is that segments of the molecule become part of a small crystal structure, known as a *crystallite*. Other segments of the molecule also pass through amorphous regions with which the crystallites are enmeshed. This tends to result in a *polycrystalline system* giving rise to a cluster or aggregate

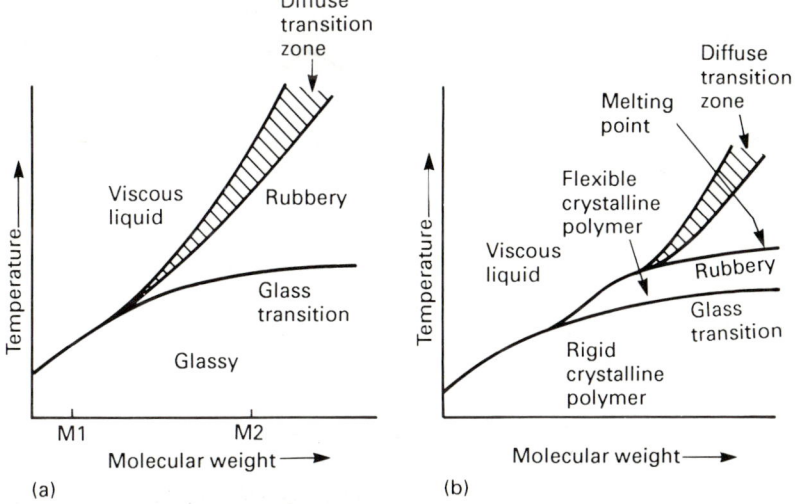

Figure 1.1 *Temperature – molecular weight diagrams for (a) amorphous and (b) moderately crystalline polymers (with highly crystalline polymers the glass transition is less apparent). From Brydson, Plastics Materials, p. 39, Figure 3.1*

of the crystallites with the overall mass of the material either having no definite shape or one imposed by a processing operation such as moulding. While differing processing conditions can give rise to different crystalline forms, polymer crystallinity can lead to the following effects important to the technologist:

(1) Where segments of molecules crystallize the molecules pack more closely together, i.e. there is an increase in density or specific gravity. Thus the crystalline polymers polyethylene and polypropylene can have densities as high as 0.96 and 0.91 g/cc, whereas the amorphous ethylene–propylene copolymer rubbers have densities of the order of 0.86 g/cc.

(2) Since in the molten state there is usually no molecular order with both amorphous and crystallizable polymers the crystalline polymers tend to shrink more on cooling, showing for example much higher levels of moulding shrinkage. (In a typical amorphous polymer this is usually about 0.005 cm/cm while with typical crystalline polymers the figure is usually in the range 0.010–0.060 cm/cm.)

(3) If a crystalline polymer is heated, a temperature will eventually be reached where the last traces of crystallinity disappear. This is known as the crystalline melting point, T_m.

(4) Any crystallization that occurs on cooling must take place between T_m and T_g, the rate varying with temperature as indicated in the diagram (Figure 1.2).

(5) Rapid cooling (*quenching*) of a product through both T_m and T_g may result in little crystallinity and hence lower shrinkage.

(6) Because the amorphous and crystalline zones usually have different densities they also have differing refractive indices (a measure of the extent a light ray is bent at a boundary or interface). Thus at each boundary between an amorphous and a crystalline zone, light rays will be deflected and *en masse* the material will appear translucent or opaque. [It may be noted that the difference in refractive indices of the two zones in polypropylene is less than with polyethylene and for this reason the former polymer is less opaque in bulk than the latter. In certain copolymers based on poly-4-methylpentene-1 (marketed as TPX polymers) the amorphous and crystalline zones have the same density and the polymer may be transparent.]

(7) As a rough guide the ratio of the glass transition temperature to the crystalline melting point is roughly given when expressed in degrees kelvin by the expression

$$T_g = {}^{2}/_{3} (T_m).$$

(8) Between T_g and T_m a thermoplastic material will range from being rubbery, through leatherlike, to hard and rigid according to the level of crystallinity present. A temperature–molecular weight diagram for a crystalline polymer will have the form shown in Figure 1.1(b).

(9) The level of crystallinity is controlled, not only by processing conditions, but also by modification of polymer structure. Thus, introduction of a small amount of a second monomer into the polymerization reactor will produce a copolymer of somewhat less regular structure than the homopolymer and thus less able to crystallize. Introduction of side groups and side chains in an irregular manner onto the backbone of the polymer molecule will have similar effects.

(10) Since, for a polymer of given molecular weight, the viscosity of a melt depends on the difference between the temperature at which the viscosity is measured and the T_g and because it is necessary to raise the temperature above the T_m to melt the polymer, crystallizable polymers tend to have lower melt viscosities than the amorphous thermoplastics at their processing temperatures. (For example: because of problems of thermal stability both PVC and polymethyl methacrylate are usually processed at less than 100 degrees Celsius above their T_g's and are viscous melts, whereas nylon

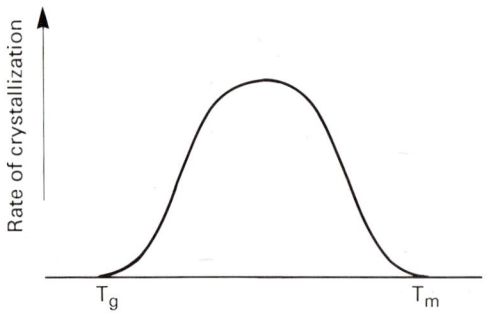

Figure 1.2 *Variations of rate of crystallization with temperature. When the T_g is only a little below normal ambient temperature after-shrinkage following injection moulding may continue for several months unless the mouldings are annealed at a temperature roughly midway between T_g and T_m at which the rate of crystallization is high*

66, which has to be processed nearly 300 degrees above its T_g is usually a fluid melt.

MOLECULAR ORIENTATION

It is important for the technologist to distinguish between crystallization and *orientation*, a phenomenon that may occur with both amorphous and crystalline polymers.

If an amorphous polymer is heated above its glass transition temperature the individual molecules tend to coil up in a random way. The chains show no particular alignment, the properties of the material are the same in every direction and the material is said to be *isotropic*. This isotropy will be largely retained on hardening of the polymer by cooling.

If, however, the polymer is subject to deformation, as during normal processing, the polymer chains tend to become uncoiled and there is a rough alignment of the molecules in the direction of the deforming stress. An example of this is where a rod of polystyrene is softened, then rapidly stretched and quickly cooled. The molecules tend to be aligned in the direction of the axis of the filament formed. This is known as *monoaxial orientation of an amorphous polymer*.

In the case where a sheet of material is stretched in two directions at once, a practice common in film manufacture, the molecules tend to become layered and often give the sheet very high values for impact strength. This is known as *biaxial orientation of an amorphous polymer*. Triaxial orientation, simultaneous stretching in three perpendicular directions, is not a practical proposition because of the very high resistance of polymers to volume change.

In the case of crystalline polymers, orientation effects are more complicated because of the orientation of crystal structures as well as of molecules in the amorphous zone. Monoaxial orientation of crystalline polymers is very important in the manufacture of synthetic fibres such as the nylons and polyester fibres. Biaxial orientation is of importance in the manufacture of polyester films and bottles. The main types of orientation are indicated schematically in Figure 1.3.

During processes such as extrusion and injection moulding, frozen-in orientation is com-

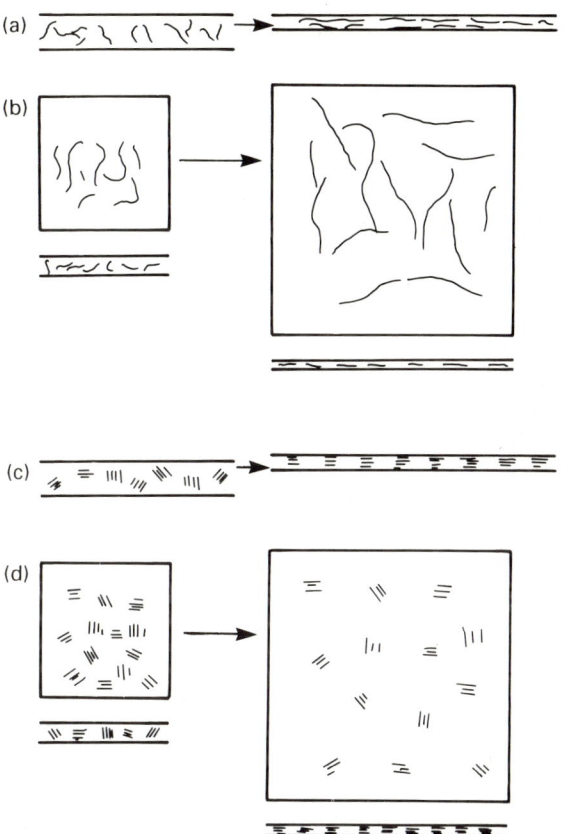

Figure 1.3 *Orientation in polymers. (a) Monoaxial orientation of an amorphous polymer. (b) Biaxial orientation of an amorphous polymer. (c) Monoaxial orientation of a crystalline polymer. (d) Biaxial orientation of a crystalline polymer*

monplace. Its main effects are to increase strength in the direction of orientation and to reduce it in directions perpendicular to the orientation. The reasons for this may be seen by reference to Figure 1.4. The top two diagrams

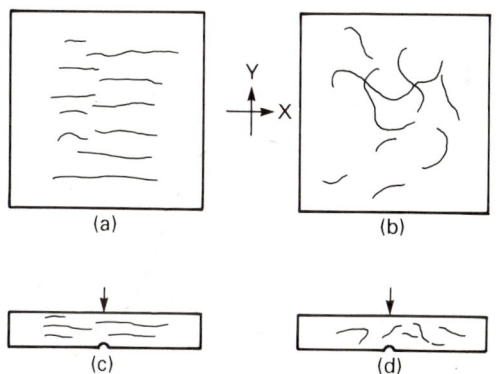

Figure 1.4 *Effect of orientation on impact resistance. The highly oriented plaque moulding struck on its face will have a lower impact resistance than the unoriented plaque. In the case of the notched bars, the oriented sample will have a higher impact strength when struck opposite the notch*

HANDBOOK FOR PLASTICS PROCESSORS

indicate two plate mouldings the first of which is highly oriented and the second free from orientation. Breaking in the X-direction in the highly oriented moulding will require rupture of molecules, whereas in the Y-direction only intermolecular forces need to be overcome. On impact onto the face of the moulding it will tend to split along the direction of orientation. There is no such preferential line in the oriented sample.

The bottom two diagrams indicate a moulded bar with a central notch. In this case striking the bar opposite the notch will require extensive molecular rupture if the highly oriented moulding is to break, but less so in the case of an unoriented moulding. In this case then, the *oriented* moulding will have the highest impact strength. Orientation, therefore, may or may not be a desirable attribute, in terms of its effect on toughness, according to the shape of the moulding.

In the case of extrusion of pipe, orientation in the direction of extrusion can make a pipe weak when subjected to hoop stresses and therefore should be minimized. Biaxial orientation may however lead to stronger extrudates.

ADDITIVES IN POLYMERS

Polymers are seldom used alone as plastics materials but are usually blended with additives. These include:

1 Inert particulate fillers (e.g. whiting and china clay used for example in PVC).
2 Reinforcing particulate fillers (e.g. carbon black and fine silicas which are used mainly in rubbers to enhance tear strength, abrasion resistance and modulus).
3 Fibrous fillers (e.g. glass fibres used widely in engineering thermoplastics and also carbon fibre).
4 Antioxidants to reduce the effects of oxygen on ageing and at elevated temperatures. These are widely used for example in polyethylene and polypropylene.
5 Stabilizers. These take many forms but of particular importance are the additives used to reduce degradation rates in PVC.

6 Plasticizers. These may make a polymer mass flexible as in the case of plasticized PVC.
7 Fire retardants.
8 Pigments.
9 Lubricants. There are many types of lubricant. *External lubricants* help to prevent sticking of molten polymer to processing equipment. *Internal lubricants* are designed to aid flow without having a plasticizing effect whilst materials such as graphite and molybdenum disulphide are used to reduce friction of the polymer against other materials.
10 Cross-linking systems – used with thermosetting plastics.
11 Blowing agents – used to make cellular, e.g. foamed, plastics.

That the use of additives may have profound effects on the properties of plastics is well demonstrated by the variety of products that may be made from PVC such as drain pipes, domestic flex, gramophone records, baby pants and playballs.

POLYMER BLENDS

Although blends of polymers have been in use for a considerable time, the range of commercial plastics of this type has increased substantially in recent years. Important examples of blends include, toughened polystyrene (usually a blend of polystyrene and polybutadiene), ABS plastics, blends of a polyphenylene oxide derivative with polystyrene (marketed as Noryl) and the thermoplastic polyolefin rubbers. In most cases the blend consists of a rigid polymer together with a more flexible polymer with the aim of combining toughness and rigidity. A number of recent blends involve a mixture of a rigid amorphous polymer with a crystalline polymer.

In principle, a blend is a physical mixture of different polymers (for example a mixture of polyethylene and polypropylene). It is thus distinct from a copolymer where the different monomers may be found on the same polymer chain (as in an ethylene–propylene rubber). A blend may be expected to have quite different properties to a copolymer. (In practice the situation may be more complicated with some blends being blends involving copolymers whilst in some instances there may be some chemical interaction between the components of a blend.)

COMPOSITES

Composite materials are widely used in the plastics industry. Indeed at one time the word *composition* was sometimes used to denote materials that would now be classified under the heading of plastics.

Composites consist of two or more physically distinct different materials which are combined in a controlled way. Three classes of composite may be distinguished:

1 Where one material (component) forms a continuous phase (*continuous matrix*) and the other components are embedded within this matrix.
2 Laminar structures where the components are in separate layers (alternating where there are just two components).
3 *Interpenetrating systems* where both components form a continuous matrix.

Further discussion will be confined to the first of these types and more specifically to *structural composites* which are characterized by the matrix material being enhanced in stiffness and strength by the presence of a *reinforcing material* embedded within it.

The continuous phase may be

(a) Metallic. Metals usually have a higher density than alternative materials and may also suffer from such problems as galvanic corrosion.

(b) Cements and inorganic glasses. These are often strong in comparison but usually have low tensile strengths. An early example of this type of composite is the insertion of wire grids into concrete to give reinforced concrete.

(c) Polymers. These are of considerable importance; the rest of this chapter deals with *polymer-based structural composites*.

POLYMER-BASED STRUCTURAL COMPOSITES

MATRIX POLYMERS

Thermosetting plastics
Whilst most thermosetting materials are available in reinforced fibre form those of most importance in making *laminates* are

Unsaturated polyester resins: These may be processed at ambient temperatures and pressures and are lower in cost than most other thermosetting resins. They are widely used in conjunction with glass fibres for car bodies, boat hulls, truck cabs and other large mouldings.

Epoxide (epoxy) resins: A wide range of resins and curing systems are available allowing a wide spectrum of properties. Appropriately formulated combinations can give laminates of higher strength and/or heat and chemical resistance than polyester resins.

Phenol-formaldehyde (phenolic) resins: These have been available for many years but usually involve curing under pressure at elevated temperatures and are therefore less versatile than polyester and epoxide resins. Fabric laminates are used for making engineering parts such as gear wheels while paper-based laminates are used in electrical applications and for the core layers of decorative laminates.

Melamine-formaldehyde resins. Although they are occasionally used with glass fibre their main use has been to provide the hard wearing colourless surfacing layers for decorative laminates.

Thermoplastics
As with thermosetting resins, most thermoplastics, and particularly those used for engineering applications, are available in grades reinforced with glass fibre. Particular reference may be made to the polyamides, polyacetals, polycarbonates, polysulphones, polyarylates and polyphenylene sulphides. The presence of 20–50% of short fibres (0.2–0.4 mm in length) leads to improved stiffness and in some cases higher heat distortion temperature. Two recent trends with thermoplastics may be noted:

1 Long fibre reinforced thermoplastics which are produced by a pultrusion process giving

longer pellets but with a modal fibre length of about 8 mm. These materials have enhanced stiffness and impact resistance. When impact failure does occur it is usually by cracking rather than complete fracture.

2 Use of heat resistant thermoplastics with carbon fibres. Of particular interest are the polyimides, polyether-imides, polyamide-imides and, in particular, carbon fibre reinforced polyether ether ketones. This latter material shows outstanding fire resistance with negligible smoke generation, very high strength and stiffness and good heat resistance. Materials of this type are of particular interest in aerospace and submarine applications.

REINFORCING MATERIAL

The *reinforcing* material may be *particulate* or *fibrous*. Particulate reinforcing materials, such as carbon blacks, are only of value with elastomers (rubbers); for non-elastomeric systems fibres are of dominant importance.

Many types of reinforcing fibres are available, the main classes being listed below.

1 Glass fibres. These are very widely used in a variety of polymers. A number of types of glass is available but two main types are recognized:

E-glass (a low alkaline aluminium borosilicate glass), fibres from which are the staple product of the glass fibre moulding industry and widely used for automotive and marine applications.

S-glass (a magnesium aluminium silicate glass), fibres from which are stronger and used, for example, in pressure bottles, rocket motor cases and missile shells.

2 Polymeric fibres. Many types have been used but two types should be mentioned here:

Cellulose fibres. These have been used for many years and include such diverse materials as paper and cotton fabric for laminates and α-cellulose and wood flour for moulding powders. In laminate form they are useful because of their high stiffness.

Aramid (aromatic polyamide) fibres. These materials are similar to glass fibres in strength but have twice the stiffness and half the density. The fibres are strong in tension

but relatively weak in compression. Composites have exceptional creep resistance and better resistance to fatigue than corresponding glass fibre composites.

3 Carbon fibres. These materials have exceptional strength and are most important in high performance structural laminates.

4 Inorganic fibres such as boron fibre and silicon carbide fibre which are of restricted application.

5 Metal wire filaments. These are probably of greatest interest in car and truck tyres.

GLASS FIBRE PRODUCTS USED FOR COMPOSITE MANUFACTURE

An individual glass fibre filament may have a diameter as little as 13 μm (0.0004 in). To protect the strands from fracture during processing and to improve bonding to polymers (and thus improve the strength of a composite) the fibre is treated with a forming size which is applied shortly after the fibres are drawn. This size will have four components:

1 A linking agent, usually a silane or sometimes a chrome complex;
2 A film former such as polyvinyl acetate which coats and protects bundles of fibres;
3 A lubricant to aid handling of fibres over guide points;
4 An antistatic agent.

Glass fibre is used in the plastics industry in a variety of forms. These include:

1 Continuous rovings. This consists of several strands, each containing hundreds of filaments gathered into bundles without twisting. Rovings may be used in filament winding and pultrusion techniques and for production of other glass products. Chopped rovings are used to make most types of glass-filled thermoplastic moulding materials.

2 Chopped strand mat. This material is the major type of reinforcement in the all-important hand lay-up technique. It consists of bundles of chopped strands, each about 5 cm in length deposited in the form of a mat

and loosely held together by a resinous (e.g. polyvinyl acetate) binder.

3 Continuous strand mat. In this case the strands are not chopped.

4 Woven rovings. Although these give laminates with high tensile strengths interlaminar adhesion is poor.

5 Woven glass cloth. The cloth may be continuous filament or from spun staple fibre. Good tensile strengths may be achieved. The cloth need not necessarily be square woven although this is the more common arrangement.

STRENGTH AND RIGIDITY OF FIBROUS COMPOSITES

The study of the strength and rigidity of composite materials is one of great complexity. This section can only point out some of the most salient features.

Let us consider the hypothetical situation of a few fibres embedded in a polymer matrix all aligned in the same direction. When stressed in tension in the direction of the fibre axes the stress will be taken up largely by the fibres and this will enhance rigidity and strength. As an approximation we may assume that if the volume fraction of fibres is V_f and the fibre modulus is E_f then in the axial direction x_1 the tensile modulus E_1 will be given by

$$E_1 \simeq V_f E_f.$$

In the direction transverse to the fibre axes x_2

$$E_2 \simeq E_p,$$

where E_p is the modulus of the polymer matrix.

Because $E_f \gg E_p$ it follows that

$$E_1 \gg E_2.$$

It is of interest to note that the shear modulus is greatest at 45° to the fibre axes.

One feature of a unidirectional system is that if a direct stress is applied to the structure at an angle to the fibres then, unlike an isotropic material, the structure will not only stretch in tension but will also distort in shear (Figure 1.5).

The strength of unidirectional systems is too dependent on the angle between the applied stress and the fibre axes to be of more than limited use.

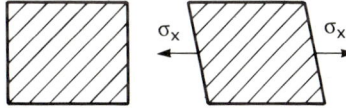

Figure 1.5 *Application of a tensile stress σ_x at an angle to the fibrous reinforcement in a laminate leads to shear strain distortion as well as a tensile strain extension. Copied from P. C. Powell, Engineering with Polymers, p. 164*

More balanced systems may be made by plying up layers of unidirectional laminae. Where the fibre axes are not at 90° to each other it is usual to refer to *angle ply*; when at 90° to *cross ply*. If such a plied laminate is not symmetrical about its mid plane (unsymmetrical laminate) then under in plane loads distortion and twisting may occur.

Laminates using randomly laid chopped strand mat are isotropic in the plane of the laminate. However because of their low fibre content they have a somewhat low stiffness and strength in the plane of the laminate.

Finally mention should be made of the fact that traditional laminates based on two-dimensional reinforcements such as chopped strand mat have limited transverse toughness and interlaminar shear strength which may lead to delamination. This has led to such developments as the use of interlocking plies.

SURVEY OF MAJOR PLASTICS MATERIALS

There are today on the market, several hundred types of plastics materials ranging from large tonnage materials such as polyethylene and PVC, to highly specialized materials such as polyvinyl carbazole and polyvinyl alcohol. In turn, with each type of material there are many grades; in the case of the major tonnage materials there may be several hundred grades for each material. It is quite clearly impossible to cover all the grades of all the materials that are available.

It has therefore been necessary to be selective, and the materials covered in this section comprise the major tonnage materials, a selection of engineering thermoplastics materials and four thermosetting materials. Whilst an attempt has been made to give typical data for a type of material, the numerical data in particular should be used with caution. Rather more important is the information given on key application properties and key processing features.

The symbols

shown in several of the chemical formulae are shorthand for ring structure of the types

respectively.

For further information on materials the reader is referred to:

Brydson J. A. (1989). *Plastics Materials*, 5th edn. London: Butterworths. (This reviews the wide range of plastics materials that have become commercially available over the years and some related laboratory materials.

Whelan A. (1982). *Injection Moulding Materials*. London: Applied Science.

Standards of Performance Manuals produced by the Plastics Processing Industry Training Board. Manuals produced to date include those on Injection Moulding, Extrusion, Blow Moulding and Thermoforming.

Cogswell F. N. (1981). *Polymer Melt Rheology*. London: Longmans.

(This contains useful appendices giving thermal, compressibility and flow data for a limited selection of polymers.)

Also of interest are a number of small privately published pocket books by Tony Whelan and John Goff. These have made their appearance during the period of preparation of this book. Three of particular note are *Injection Moulding of Engineering Thermoplastics, Moulding of Thermosetting Plastics*, and *Injection Moulding of Thermoplastics Materials*, 1 and 2. They are available from the authors or from the Plastics and Rubber Institute.

Materials listed

ABS POLYMERS

Alternative names
Acrylonitrile-butadiene-styrene
ABS Copolymers
ABS Terpolymers

Polymer structure
These materials are complex blends and there are many variations. In most types acrylonitrile and styrene are grafted onto a polybutadiene backbone. The product may also contain unreacted polybutadiene and some acrylonitrile-styrene copolymer.

Polymer status
Very widely used polymer for 'engineering applications' which do not have severe load bearing demands. Particularly important for housings (both domestic and industrial) and in the automotive industry.

Principal variants
Grades vary in molecular weight and in the acrylonitrile–butadiene–styrene ratio. Blends with polycarbonates and polysulphones are also of importance.

Key application properties
Toughness, rigidity and good surface appearance of well-moulded products. Also generally better heat and chemical resistance than polystyrene.

Selected typical properties

Specific gravity	1.01–1.07
Tensile strength (MPa)	31–45
Modulus (MPa)	1515–2620
Elongation at break	Very dependent on formulation and extension rate
Impact strength (Izod) ASTM D256	1–8ft lbf in^{-1} notch
Glass transition temperature	Not really applicable to a blend of this type
Crystalline melting point	Amorphous. No melting point
Deflection temperature under load (°C)	
(1.82MPa)	87–107
(0.48MPa)	90–107
Coefficient of thermal expansion	
(cm/cm/°C)	7.9–11 × 10^{-5}
Volume resistivity (Ωcm)	3 × 10^{15}
Power factor	0,007–0.31

Chemical and solvent resistance Generally resistant to alkalis and acids but not concentrated oxidizing acids. Dissolved by many aromatic and chlorinated hydrocarbons, esters and ketones.

Maximum service temperature UL temperature index 60–75 °C. The very high impact grades tend to be at the lower end of this range. Alloys with polycarbonate may be as high as 95 °C.

Burning properties Standard grades are considered as slow burning and usually meet the UL HB requirement. The material burns with a smoky yellow flame emitting a pungent gas. Fire retardant grades are not wholly satisfactory and where flame retardance is important it may be better to consider ABS/PVC alloys.

Key processing properties
- May absorb up to 0.3% water and therefore must be stored under dry conditions.
- Greater tendency to degradation than polystyrene during processing so important not to overheat. Avoid too high a screw rotational speed and back pressure during moulding.
- Generally less free flowing than polystyrene, particularly with heat resistance grades. Flow path ratios 80–150:1
- Low moulding shrinkage typical of an amorphous polymer (0.004–0.008 cm/cm).

Typical injection moulding conditions

Recommended melt temperatures (°C) 220–280 according to type
Temperature settings (°C)

Cylinder – rear	170–220
– middle	180–225
– front	220–270
Nozzle	210–270
Mould	40–90

(The higher values are more relevant to heat resistant grades)
Injection pressures (MPa)

1st stage	up to 150
2nd stage	up to 75

Shutting down and reprocessing Shut off heat and purge clean. To reduce the level of fumes in the atmosphere purgings should be immersed in an earthed container of cold water.

Other comments: A high quality finish is an important requirement for many mouldings. This will require dried material, good temperature control and, if available, profiled injection.

Typical extrusion conditions

Melt temperature (°C)	205–250
Temperature profile (°C)	190–250
Recommended screw L/D ratio	20:1 to 36:1
Recommended compression ratio	2,5:1 to 3:1

Other comments: It is important that dies are streamlined and that the melt not be allowed to stagnate and decompose.

Thermoforming

ABS may be formed by vacuum forming, pressure forming and mechanical forming methods. In the case of vacuum forming a good vacuum is necessary because of somewhat high elastic modulus of the material in the rubbery state required for forming. It is essential that the sheet is dry before attempting to thermoform.

Forming temperatures (°C)	130–190
Set temperature (°C)	85

Applications

Widely used where toughness, rigidity and good appearance are of importance and where these may be achieved with reasonable ease of processing and moderate cost. Uses include automotive components (instrument panels, armrests, grilles), appliance housings, luggage, business machine housings, garden equipment, caravan and boat panels.

ACETAL POLYMERS

Alternative names
Polyacetal
Polyformaldehyde
Polyoxymethylene

Polymer structure

$$-CH_2-O-$$ Homopolymer

$$+CH_2-O+-CH_2-CH_2-O+$$ Copolymer
c. 98%   c.2%

Polymer status
Major engineering thermoplastics material usually processed by injection moulding.

Principal variants
There are two main types, homopolymers and copolymers. The homopolymers have a more regular structure and are therefore rather more crystalline and exhibit slightly higher density, strength, stiffness and softening point. Copolymers tend to have slightly better stability and mouldability. Modified grades include glass filled-, carbon fibre filled-, ptfe filled- and rubber modified.

Key application properties
Generally considered superior to the nylons in fatigue endurance, stiffness, creep resistance and water resistance, but inferior in impact toughness and abrasion resistance. Low coefficient of friction (0.1–0.3).

Selected typical properties

Specific gravity	1.425 (homopolymer); 1.410 (copolymer)
Tensile strength (MPa)	70 (homopolymer); 58 (copolymer)
Modulus (MPa)	2800 (homopolymer); 2500 (copolymer)
Elongation at break	15–75%
Impact strength (Izod) (ASTM D256)	1.1–2.3 ft lbf in^{-1} notch
Glass transition temperature	−13 °C
Crystalline melting point (°C)	175 (homopolymer); 163 (copolymer)
Deflection temperature under load (°C)	
(1.82 MPa)	100–110
(0.48 MPa)	158–170
Coefficient of thermal expansion	
(cm/cm/°C)	$8.1–8.5 \times 10^{-5}$
Volume resistivity (Ωcm)	$10^{15}–10^{16}$
Power factor (100–10 000 Hz)	0.004

Chemical and solvent resistance
Acetals show very good resistance to organic chemicals with no solvents below 70 °C although some with similar solubility parameter may cause swelling. Not resistant to acids and bases (although copolymer has much better resistance to bases). Attacked by dilute mineral acids and by strong organic acids.

Thermal stability
UL thermal index (mechanical without impact) 90 °C

Flammability
Classified as slow burning. It drips as it burns with a smokeless blue flame.

Key processing properties
- Although less hygroscopic than nylons (<1% water absorption) should be stored in a dry area. In such circumstances predrying is not normal but where necessary typically 2–3 h at 110 °C.
- Injection moulding grades flow easily although polymers available with range of flow properties (usually evaluated by use of the *melt flow index*). For general purpose work MFI 9 common, but for high precision and/or complex mouldings lower molecular weight grades of easier flow (MFI 27) may be used. Flow path ratios 100–250:1
- Acetals have limited thermal stability. Do not exceed suppliers' recommendations on maximum temperatures and times at elevated temperatures. Never mix with PVC compounds since PVC degradation products accelerate decomposition (sometimes explosively) to monomer.
- Because polymers are highly crystalline moulding shrinkage is high (0.2–0.035 cm/cm). Reinforced grades exhibit differential shrinkage. Some after-shrinkage (up to 0.1%) may occur during 48 h immediately after moulding.

Typical injection moulding conditions

Recommended melt temperatures (°C)	205 (homopolymer); 185 (copolymer)
Temperature settings (°C)	
Cylinder – rear	165–170
– middle	170–200
– front	180–210
Nozzle	170–210
Mould	40–120

Injection pressures: Up to 170 MPa for both first and second stage.

Shutting down and reprocessing Purge clean and leave screw in forward position

Other comments
Screw rotational speeds may be up to 200 r.p.m. Back pressure typically 20 MPa but may be much lower.

Typical extrusion conditions

Temperature settings (°C)	170–190
Recommended screw L:D ratio	20:1
Recommended compression ratio	3:1–4:1

Other comments: Take care against degradation which occurs rapidly above 200 °C. Always purge with polyethylene. Immerse degraded extrudate materials immediately in water.

Applications
Acetal polymers have been widely used as metal replacements for such diverse uses as handles, switch parts, cams and gears, pump impellers, machine linkages and couplings, plumbing components, water jugs, zippers and aerosol containers.

AMINOPLASTICS

Alternative names
This section covers two similar thermosetting materials:

    Urea-formaldehyde plastics (UFs)

    Melamine-formaldehyde plastics (MFs)

These two materials are often referred to respectively simply as *ureas* and as *melamines*. This is not to be encouraged as there can be confusion between the two plastics with the chemicals urea and melamine.

Polymer structure
Highly complex networks based on carbon, oxygen, nitrogen and hydrogen.

Polymer status
UF, and to less extent the more expensive MF, resins are used in a wide range of applications such as plywood and chipboard manufacture, wood and paper adhesives, manufacture of wet strength paper, surface coatings, and thermosetting moulding materials.

UF moulding materials continue to be widely used in the electrical industry for plugs, sockets, switches, connectors and lamp fittings. They also find such varied uses as buttons, bottle caps and toilet seats.

MF moulding materials are mainly known for tableware but also find some use for other rigid, heat resistant uses which demand good finish and bright colours.

Principal variants
Besides the two basic types (UF and MF) grades also differ in the type of filler. α-cellulose fillers are widely used with wood flour also being used in cheaper UF grades. Glass fibre filled MF resins have some specialist electrical uses.

Key application properties
UF materials are attractive because of their rigidity (as with PFs and other thermosetting materials) but unlike PFs are available in a wide colour range, do not impart taste and odour to foodstuffs with which they come into contact and have good electrical properties with particularly good resistance to tracking. They have low cost.

Disadvantages include limited heat resistance (continuous exposure is limited to temperatures below 70 °C), poor scratch resistance, high water absorption and poor staining resistance.

MF materials have similar advantages over the PFs and additionally have better heat resistance, water resistance, scratch resistance and strain resistance compared with the UFs. They are more expensive.

Selected typical properties

	UF GP	MF GP	MF Glass filled
Specific gravity	1.5–1.6	1.5–1.55	c.2.0
Tensile strength (MPa)	52–80	55–83	41–69
Modulus (MPa)	9000–11000	1100–16500	–
Rockwell hardness (M)	110–120	115–125	–
Elongation at break (%)	< 1	1.6–2.4	–
Impact Strength (ft lbf in^{-1})	0.2–0.35	0.15–0.24	0.16–0.23
Glass transition temperature	Material is thermoset		
Crystalline melting point	Material is thermoset and amorphous		
Deflection temperature under load (°C)			
(1.82 MPa)	125–145	175–200	–
Coefficient of thermal expansion			
(cm/cm/°C)	$22–36 \times 10^{-6}$	$40–45 \times 10^{-6}$	–
Volume resistivity (Ωm)	$10^{13}–10^{15}$	$10^9–10^{10}$	$10^{15}–10^{16}$

Chemical and solvent resistance
Since the mouldings are cross-linked there are no solvents. Good resistance to detergents and dry cleaning chemicals leads to uses in clothing. Good resistance to weak alkalis but limited resistance to acids and strong alkalis.

Water absorption of UFs is quite high (50–130 mg in 24 h at 20 °C using the BS2782 test.) MFs are much better (10–50 mg for α-cellulose-filled and 10–20 mg for glass-filled). Poor staining resistance.

Maximum service temperature About 70 °C for continuous use.

Burning properties Material has good resistance to burning with a V-O UL94 rating and an oxygen index of 35%.

Key processing properties
- Moulding powders are somewhat hygroscopic and should be stored in a cool

dry place, preferably in sealed containers (particularly important with UFs).

- Moulding powders have a limited shelf life tending to cure slowly at room temperature. UFs have shorter shelf life (*c.* 6 months). MF materials should last at least one year.

- Most grades of MF and UF are easy or medium flow, but flow of old UF powders is more difficult. Stiff flow grades may be preferred for deep-draw articles. UFs tend to evolve more gases on cure than MFs so that bumping (briefly reopening the mould just after closure) is often essential in compression moulding.

- Moulding shrinkage is usually at the order of 0.5–0.8%.

Typical injection moulding conditions

	UF	MF
Recommended melt temperatures (°C)	120–140	120–140
Temperature settings (°C)		
Cylinder	65–95	85–105
Nozzle	85–110	95–120
Mould	135–165	145–180
Injection pressures (MPa)	100–250	100–250
Cure Time (s)	15–80	15–80

Other comments: Degree of cure of UF may be simply assessed by noting effect of boiling a cut section in water for 10 min. There should be no visual deterioration and the sample should not be scratched with a fingernail. In the case of MF mouldings sample may be boiled in a 0.01% aqueous solution of Rhodamine B. Apart from any cut edges and parting line the moulding should not be stained. For some applications it may be necessary to use more specific tests since different properties change with cure at different rates and the above tests only provide a compromise.

Typical compression moulding conditions

	UF	MF
Preheat temperatures (°C)	95	100
Mould temperature (°C)	135–165	140–180
Moulding pressure (MPa)	15–60	15–60
Cure Time (s)	30–180	40–210

Typical transfer moulding conditions

	UF	MF
Mould temperature (°C)	150–170	150–170
Transfer pressure (MPa)	60–120	60–120
Cure time (s)	30–120	30–120

Applications

UF moulding materials used for electrical fittings (plugs, sockets etc), bottle caps, toilet seats. MF materials used primarily for tableware.

DOUGH MOULDING COMPOUND

Alternative names
DMC
Bulk moulding compound
Polyester moulding compound

Polymer structure
These compounds are blends of complex polyester resins with glass fibre, fillers, curing agents and other additives.

Polymer status
An important thermosetting material because of key properties listed below.

Principal variants
A variety of polyester resins may be used. Whereas most are based on phthalic acid, resins based on isophthalic acid are used to give enhanced heat resistance. Low profile resins, with a smoother surface may be obtained by the addition of certain thermoplastic materials. Variants may include additives to enhance toughness and flame resistance.

Key application properties
High flexural modulus. Low coefficient of thermal expansion. Very good heat resistance (up to 200 °C) and corrosion resistance. Moderate cost.

Selected typical properties

Specific gravity	1.7–2.1
Tensile strength (MPa)	10–20
Modulus (MPa)	3400–4500
Impact strength (ft lb in^{-1}) (BS771)	2–4
Glass transition temperature	–
Crystalline setting point	Amorphous thermosetting material
Volume resistivity (Ωcm)	> 10^{16}
Power factor	0.02–0.08

Chemical and solvent resistance Does not dissolve in any solvents. Limited resistance to ketones and chlorinated hydrocarbons. Resistance to aromatic hydrocarbons, alkalis, and acids is also not very good, particularly at elevated temperatures.

Maximum service temperature 160 °C for long periods: 200 °C for intermittent exposure.

Burning properties Unmodified grades will burn but flame retardant grades may be rated as UL94 V-O.

Key processing properties
- DMC has low water absorption and is easy flow
- Thermal stability is very good.
- Shrinkage is very low and negligible with some grades.
- Material has limited storage life (e.g. 3 months)

Typical injection moulding conditions

Temperature settings (°C)		
Cylinder		20–60
Nozzle		20–40
Mould		140–160
Injection pressures (MPa)	25–80	
Curing time (s)	20–90	

Other comments: DMC is widely used for cookware (particularly microwave). It is important to remove traces of styrene or other monomer present in the formulation before use as this can be tasted in the food. One procedure is to bake for 15 min at 180 °C. The container is then cooled, filled with water and then heated again to leach out any residual styrene. During processing keep working area well-ventilated.

Typical compression moulding conditions

Preheat temperature (°C)	80
Mould temperature (°C)	140–150
Moulding pressure (MPa)	3.5–15
Cure time (s)	25 (1 mm thick): 45 (2.5 mm thick)

Applications
Microwave and mid-temperature range oven cookware. Domestic iron and toaster components. Office machinery. Car headlight reflectors. Electrical meter boxes and lids. Safety helmets. SMC (Sheet Moulding Compound) is a closely related product.

PHENOLIC PLASTICS

Alternative names

Phenol-formaldehyde plastics. The term phenoplasts has been used in the past but is now archaic. At one time the materials were also probably better known under the trade name Bakelite although today there are many other manufacturers and also the trade name is also used for other plastics. They are also widely known simply as PFs.

Polymer structure

The structure is highly complex and irregular with the polymer cross-linked after the shaping stage. A typical segment structure could be of the form

Polymer status

Less important today than 25 years ago. Laminated phenolics are still of great importance but moulding materials are now largely restricted to applications demanding moderate levels of heat resistance.

Principal variants

The following comments apply only to moulding materials.

Standard moulding materials are based on using resin and wood flour in approximately equal amounts, the wood flour both improving impact strength and reducing cost. Cotton flock or chopped fabric instead of wood flour increases toughness whilst the use of glass fibre will improve rigidity and dimensional stability. Whilst most grades are

based on phenol, improved acid resistance will be obtained by the use of cresols and improved alkali resistance by the use of xylenols. Furfural has been used in the past instead of formaldehyde. Melamine-phenol-formaldehyde moulding materials exhibit many of the desirable properties of phenolics together with the ability to be produced in a wide range of colours. There are two main types of phenolic resin depending largely on the phenol:aldehyde ratio and the reaction conditions during resin manufacture. These are the *novolaks*, which are the normal type used for moulding materials, and *resols*, which occasionally find use where minimum odour, and/or enhanced alkali resistance is required.

Key application properties

Good heat resistance (i.e. resistance to thermal degradation). Since products are thermoset the material does not melt. Mouldings burn only with difficulty and tend to char. High rigidity. Good water resistance although this depends on both type of resin and of filler used. Electrical insulation properties are only fair and electrical tracking resistance and related properties are poor. Because of chemical reactions occurring during cure the mouldings are limited to dark colours only.

Selected typical properties

(All properties based on BS 2782 test methods)

	General purpose	High shock	Electrical low loss	Heat resistant
Specific gravity	1.35	1.40	1.85	1.94
Tensile strength (MPa)	55	45	58	34
Modulus (MPa)	7000–15000	–	–	–
Elongation at break %	0.1–0.5	–	–	–
Impact strength (J)	0.22	1.08–1.9	0.18	0.13
Glass transition temperature (°C)	Decomposes below the T_g			
Crystalline melting point (°C)	Polymer is amorphous			
Deflection temperature under load (1.82 MPa) (°C)	←————— 150–260 ——————→			
Coefficient of thermal expansion (cm/cm/°C)	←————— 10–40×10^{-6} ——————→			
Volume resistivity (Ωcm)	10^{12}–10^{14}	10^{12}–10^{14}	10^{14}–10^{16}	10^{11}–10^{14}
Power factor (800 Hz)	0.1–0.4	0.1–0.5	0.03–0.05	0.1–0.3

Chemical and solvent resistance Chemical resistance depends on type of resin and filler used. Simple PFs are attacked by alkalis but are resistant to most acids other than 50% sulphuric acid, formic acid and oxidizing acids. Cresol- and xylenol-based resins have improved alkali resistance.

Maximum service temperature The resins are generally stable to 200 °C but many common fillers are much less stable. Some grades have been claimed to be stable up to 300 °C.

Burning properties Phenolic moulding materials burn only with difficulty; they do not drip but tend to char.

Key processing properties
- The novolak based resins have good water resistance but some fillers are somewhat hygroscopic. This is not normally a special problem as gases are in any case evolved during the curing stage and provision should be made for these to escape, either by 'bumping' (briefly reopening the mould just after closure) in the case of compression moulding or by adjusting temperature profiles in injection moulding.
- The general purpose grades are available in a range of viscosities ('flows'). The fibre-filled grades such as the high shock types do not flow easily. Fibres are less damaged by compression moulding.
- Moulding powders only deteriorate very slowly on storage. At processing temperatures cross-linking is rapid (and very dependent on temperature). It is very important to avoid premature curing in the injection unit of an injection moulding machine.
- Shrinkage of general purpose compression moulding grades is about 0.8% (with glass-filled grades down to 0.3%). Injection moulding grades range from 0.8 to 1.7%.

Typical injection moulding conditions

Recommended melt temperatures (°C)	110–140
Temperature settings (°C):	
Cylinder	65–90
Nozzle	85–120
Mould	165–195
Injection pressures (MPa)	85–250
Cure time (s)	15–80

Typical compression moulding conditions

Preheat temperature (°C)	70–105
Mould temperature (°C)	165–185
Moulding pressure (MPa)	15 (minimum)
Cure time (s)	30–60

Typical transfer moulding conditions

Mould temperature (°C)	165–185
Mould pressure (MPa)	50 (min)
Cure time (s)	40–120

Applications

Electrical applications involving low frequency and without demanding tracking conditions, particularly for elevated temperature work. Examples include coil bobbins, terminal blocks, relay bases and component housings. Domestic uses include knobs and saucepan handles, cooker handles. Glass-filled grades because of their high rigidity are often used for thin-walled electrical mouldings.

(ALIPHATIC) POLYAMIDE HOMOPOLYMERS

Alternative names
Nylons
Polyamides

Although the term polyamide is now widely used to describe the materials covered in this section this term may also be used to describe a much wider range of materials including aromatic polyamides, thermoplastic polyamide rubbers and polyamide resins used in surface coatings and with epoxide resins. For simplicity the term nylon will be used here.

Polymer structure
All polyamides have the repeating unit $-CONH-$ in their molecular structure. The different types are described below under the heading Principal Variants.

Polymer status
When selecting a thermoplastics material for light engineering applications the nylons are usually the first material to be considered and may be considered as the leading engineering thermoplastic. Of the many types the market is dominated by nylons 6 and 66.

Principal variants
Grades vary in their basic structure, in their molecular weight and in the additives that may be present. All of the nylons considered in this section are made by one of the three following routes.

1 Reaction of a diamine with a dibasic acid. For example nylon 66 (polyamide 66) is made by reacting hexamethylene diamine with adipic acid. The first 6 indicates the number of carbon atoms in the amine, the second the number in the acid. Other polyamides made this way include: nylon 46 (tetramethylene diamine and adipic acid); nylon 610 (hexamethylene diamine and sebacic acid)

2 Ring opening of a lactam. Polymers made in this way are: nylon 6 (polycaprolactam made by ring opening of the caprolactam ring which contains 6 carbon atoms); nylon 12 (made from lauryl lactam with 12 carbon atoms)

3 Self-condensation of an ω-amino acid, e.g. nylon 11 which is made from an amino acid with 11 carbon atoms.

$\sim (CH_2)_6\, NHCO\, (CH_2)_4\, CONH \sim$	nylon 66
$\sim (CH_2)_5\, CONH \sim$	nylon 6
$\sim (CH_2)_{11}\, CONH \sim$	nylon 12

As a very rough guide, the more carbon atoms between the repeat amide groups, the more the material approaches the properties of polyethylene. For example nylon 12 is intermediate in properties between nylon 6 and polyethylene; nylon 46 is even less like polyethylene than nylon 66.

Key application properties
Toughness with rigidity, abrasion resistance, good hydrocarbon resistance, fair–good heat resistance. High water absorption may be a problem, while electrical insulation properties are only fair at high frequencies.

Selected typical properties of unfilled grades

	46	66	6	610	11	12
Specific gravity	1.18	1.14	1.13	1.09	1.04	1.02
Tensile strength (MPa)	100	80	76	55	38	45
Tension modulus (MPa)	3000	3000	2800	2100	1400	1400
Elongation at break (%)	<------- 80–200 ------->					
Impact strength (Izod) (ft lbf in^{-1}) (Dry)	–	1.3	1.0	1.1		
Crystalline melting point (°C)	295	264	215	215	185	175
Deflection temperature under load						
(1.82 MPa) (°C)	160	75	60	55	55	51
(0.48 MPa) (°C)	–	200	155	160	150	140
Coeff. of thermal expansion (cm/cm/°C)	<------- 9–15×10^{-5} ----->					
Volume resistivity (Ωcm)	<-- (>10^{17} dry: c.10^{14} @ 65% RH) -->					
Power factor	<------- 0.02–0.06 ------->					

Chemical and solvent resistance Good resistance to hydrocarbons but decreasing going from left to right in this table. Alcohols may cause some swelling. Formic acid, glacial acetic acid, phenols and cresols are solvents. Types with highest hydrocarbon resistance absorb most water:

(Equilibrium water absorption @ RT   –   8%   11%   3.5% 2%   1.7%)

Maximum service temperature For continuous use 75 °C but may be used up to 150 °C for short term application. Better in absence of air or light.

Burning properties Unmodified grades will burn with melting. Suitable flame retardants are available that will confer a UL V-O classification.

It should be noted that the mechanical properties of the nylons are sensitive to both temperature and moisture content. The presence of moisture can act like a plasticizer and greatly increase the toughness. In the case of nylon 46 the impact strength is four times higher after conditioning to 62% RH than it is when dry as moulded.

The following data is typical of glass-filled nylon 66

Specific gravity	1.38
Tensile strength (MPa)	200
Elongation at break	3–5% (c. f 80–100% for unfilled polymer)
Deflection temperature (1.82 MPa)	245°C (c. f 75 °C)
(0.45 MPa)	254°C (c. f 200°C)
Coefficient of expansion (cm/cm/°C)	2.8×10^{-5}

Key processing properties
- The nylons are hygroscopic and it is essential that materials are thoroughly dry before moulding or extrusion. Where granules have become damp drying in air should be carried out at temperatures below 80 °C to reduce the tendency of granules to oxidise so that moulding may be discoloured and/or brittle. If the granules are very damp it may be necessary to dry in a vacuum for up to 12 h at about 100 °C.
- The nylon melts are very fluid. Flow path/wall thickness ratios can be as high as 340/1 for low viscosity melts; down to 140/1 for high viscosity melts. Because of the low melt viscosity and melt elasticity it is difficult to handle extrudates.
- At the high processing temperatures (a result of the high T_m's of these polymers) oxidation can occur if air comes into prolonged contact with the melt.
- Since they are crystalline materials the moulding shrinkage is much higher than for amorphous plastics such as polystyrene and varies with moulding conditions. Unfilled grades have values typically in the range 0.01–0.02 cm/cm. The shrinkage of

glass filled grades may be as low as 0.003 cm/cm in the flow direction but similar to unfilled grades in the transverse direction. Some varieties, e.g. nylon 66 exhibit after-shrinkage which may take over a year to reach equilibrium at room temperature. This may be accelerated by heating in a non-oxidizing fluid at the temperature of maximum rate of crystallization (about midway between the T_g and the T_m). Where dimensions are critical it is necessary to take into account swelling due to water absorption as well as the opposing effects of shrinkage and after-shrinkage.

Typical injection moulding conditions

	46	66	6	11/12
Recommended melt temperatures (°C)	310	280	250	260
Temperature settings (°C):				
barrel (rear)	280	280	230	225
(middle)	310	280	245	245
(front)	310	275	250	255
nozzle	250	270	240	250
mould	120	20–100	60–90	30–100
Barrel settings for reinforced grades are about 10 °C higher.				
Injection pressures (MPa)	100	100	75–150	50–100

Shutting down and reprocessing Use more viscous polyester, e.g. PP or HDPE

Typical extrusion conditions

Temperature settings. Use similar profies as with injection moulding
Recommended screw L/D ratio   15–20/1 (25/1 with vented screws)
Recommended compression ratio   8.5–4/1

Other comments: Use very short compression zone. Vented barrels very useful to remove traces of water. Special care and techniques required to handle the low viscosity/low elasticity extrudates.

Applications
Very widely used for light engineering applications. Particularly valuable where toughness and abrasion resistance are important. Unfilled grades inferior to polyacetals in stiffness, fatigue endurance, creep resistance and with a higher coefficient of friction. Today there is substantial use of glass filled grades.

POLYBUTYLENE TEREPHTHALATE

Alternative names
PBT
Polytetramethylene terephthalate
PTMT

Polymer structure

$$\sim OOC \overbrace{\bigcirc} COO \cdot CH_2 \cdot CH_2 \cdot CH \cdot CH_2 \sim$$

Polymer status
Engineering plastics material.

Principal variants
In addition to unmodified polymer:

1 Glass fibre filled grades
2 Mineral filled grades (usually in conjunction with glass fibre)
3 Flame retardant grades (with or without reinforcement)
4 Rubber-modified grades
5 Blends with other polymers, e.g. polycarbonates.

Key application properties
Superior to nylons in respect of acid resistance, lower moisture absorption and lower coefficient of friction. Inferior to nylons in respect of alkali resistance and abrasion resistance. Similar in most other respects.

Selected typical properties

	Unfilled	Reinforced
Specific gravity	1.32	
Tensile strength (MPa)	52	up to 140
Modulus (MPa)	2350	up to 10 400
Elongation at break (%)	200	2
Impact strength (ft lbf in^{-1}) notch	1.0	up to 2
Glass transition temperature (°C)	22–43	
Crystalline melting point (°C)	224	
Deflection temperature under load (°C)		
(1.82 MPa)	54	210
(0.48 MPa)	154	215
Coefficient of thermal expansion (cm/cm/°C)	1.1–5 × 10^{-5}	
Volume resistivity (Ωcm)	4 × 10^{19}	6 × 10^{19}
Loss tangent (60 Hz)	0.0014	0.0017

Chemical and solvent resistance
Good resistance to most acidic solutions but ester linkages in polymer structure are susceptible to attack (hydrolysis) by alkalis. Low water absorption at usual ambient temperatures. Good resistance to hydrocarbons, chlorinated hydrocarbons, alcohols, ketones, esters, petrol (gasoline), fuel oils.

Maximum service temperature
UL Temperature Index 105–140 °C. For unfilled grades the low deflection temperature may provide a lower limit if part used under load.

Burning properties
Both reinforced and unreinforced grades in the absence of flame retardants are rated UL94 HB. Flame retardant grades can meet UL94 V–O and in some cases UL94 5–V.

Key processing properties
• Although polymer absorbs low amount of water this can be enough to cause some hydrolysis at elevated temperatures. Water content should not exceed 200 p.p.m. Dry in an oven for 3 to 5 h from 120 to 150 °C or in a dessicant drier 2.5 h at 120 °C.
• Low melt viscosity at normal processing temperatures permits easy filling of thin wall mouldings.
• The polymer has comparatively poor thermal endurance. Avoid temperatures above 270°C to avoid thermal damage (damage by hydrolysis due to traces of water will occur at lower temperatures). Even at 200 °C there is some degradation so that barrel residence times, particularly at the higher processing temperatures should be kept as short as possible.
• The polymer shows a rapid rate of crystallization which allows short cycle times.
• Shrinkage of unfilled grades is about 0.015–0.020 cm/cm but glass fibre filled grades will be as low as 0.003–0.008 cm/cm. Shrinkage may be twice as great transverse to the flow as in flow direction and this can cause warping; this may be reduced by use of mineral filled compounds.

Typical injection moulding conditions

Recommended melt temperatures (°C)	230–260
Temperature settings (°C)	
Barrel – rear	215–230
middle	225–250
front	205–230
Nozzle	195–215
Mould	20–110

Injection pressures: Machine should be capable of giving 130 MPa for first stage and 72 MPa for the second stage.

Shutting down and reprocessing If there is a break in production, empty barrel and bring screw forward before shutting off. At end of production purge with HDPE.

Other comments: Because of thermal sensitivity of the polymer it is recommended that only 50–80% of rated capacity of machine should be used for each shot to reduce the thermal history.

Applications
Widely used in broad range of applications such as automotive, electrical and electronic, housewares, materials handling, power tools and toys.

POLYCARBONATE

Alternative names
Polycarbonate of bis-phenol A

Polymer structure

Polymer status
Widely used thermoplastics material where toughness is a prerequisite.

Principal variants
Most commercial grades are based on the homopolymer with the structure given above. There is some limited use for copolymers and for carbonates other than that based on bis-phenol A. Blends with ABS and PBT are of importance in car components and there is increasing use of glass and carbon fibre filled grades.

Key application properties
Toughness, rigidity, transparency, heat resistance, good electrical insulation characteristics and self-extinguishing. *However*, material is notch sensitive, should not be used under long term loading which may cause strains in excess of 0.75% and like many other aromatic polymers has limited resistance to electrical tracking and arcing. It is also subject to cracking and crazing in presence of certain liquids and care should be used when painting or using adhesives.

Selected typical properties

	Unfilled	Glass-filled
Specific gravity	1.2	1.3–1.45
Tensile strength (MPa)	65	110–130
Modulus (MPa)	2400	6000–10000
Elongation at break (%)	6–7	2–4.5
Impact strength (ft lbf in^{-1}) notch Izod		
(ASTM D256 $1/2''$ × $1/8''$ bar)	12–18	c.2.5
Glass transition temperature (°C)	c.145	

Crystalline melting point (°C)	220–230°C	
Deflection temperature under load (°C)		
(1.82 MPa)	135–140	149–154
(0.48 MPa)	140–146	146
Coefficient of thermal expansion (cm/cm/°C)	7×10^{-5}	1.7–2.7×10^{-5}
Volume resistivity (Ωcm)	2.1×10^{18}	$>10^{16}$
Power factor	0.009 at 60 Hz, 0.0021 at 10 kHz, 0.010 at 1 MHz	

Chemical and solvent resistance Somewhat limited resistance to hydrolysing agents. Not resistant to caustic soda, caustic potash, ammonia and many aromatic amines. Some resistance to dilute mineral acids. Dissolves in solvents of similar solubility parameter which have proton donating ability (e.g. symtetrachloroethane and dichloromethane, chloroform and 1,1,2-trichloroethane). Several liquids will cause cracking, e.g. carbon tetrachloride unless special grades are used whilst other solvents may start to dissolve the amorphous material which then may start to crystallize from solution. Good resistance to oxygen.

Maximum service temperature 100–135 °C

Burning properties Difficult to ignite and usually extinguishes on removal from flame which is yellow and sooty. Unmodified unfilled grades typically have a UL94 V–2 rating and filled grades UL94 V–1. Flame retardant grades may be rated UL94 V–O and UL94 5–V.

Key processing properties
- Tends to pick up small amount of water (up to 0.3%) but which will produce substantial amount of steam at the high processing temperatures and thus important to keep material very dry before processing. (Typical drying times 2-4 h at 120 ° C with dried granules kept in dried heated hoppers at about 80 °C.)
- Polycarbonates generally have high melt viscosities although easy flow grades are available. Melt viscosity is less dependent on temperature and shear rate than for many other polymers. Flow path: wall thickness ratios are generally quite low (30:1 to 70:1).

- Polymer has good thermal stability.
- Moulding shrinkage 0.006–0.008 cm/cm which is slightly higher than for totally amorphous materials. (Glass filled grades 0.003–0.005 cm/cm).

Typical injection moulding conditions

Recommended melt temperatures °C		280–320
Temperature settings (°C)	Unfilled	Filled
Barrel – rear	275–300	300–315
middle	285–315	310–345
front	285–315	315–330
Nozzle	280–310	310–330
Mould	80–120	70–130

Injection pressures: Machine should be capable of giving
First stage: 200 MPa (2000 bar, 29,000 p.s.i.)
Second stage: 120 MPa (1200 bar, 17,400 p.s.i.)

Shutting down and reprocessing Problems may occur if machine is switched off with barrel full. High polymer to metal adhesion may cause small metal fragments to be pulled away from barrel wall as polymer shrinks on cooling. Often possible to keep heaters on at temperatures above the T_g of the polymer (e.g. at about 170 °C) when production is interrupted. For long interruptions purge with HDPE.

Other comments: Level of frozen-in strain may be checked by immersion in carbon tetrachloride. Select moulding conditions which gives lowest level of cracking. Level of frozen-in strain may be reduced by annealing at 125 °C for up to 24 h. Annealing at temperatures above the T_g may allow crystallization to occur and the moulding may become milky or even opaque.

Typical extrusion conditions

Melt temperatures (°C)	280–310
Temperature profile °C	270–315
Recommended screw L/D ratio	20:1
Recommended compression ratio	1.5:1 to 2.5:1

Typical thermoforming conditions

Forming temperature:	180–220 °C
Set temperature:	140 °C

Applications

Widely used in electronics, electrical engineering, medical applications, glazing and tough glazing applications. Used in cameras, domestic mixers and other kitchen equipment. Audio compact discs. Blends (with PBT) used for car bumpers, front ends, wings and other components.

POLYETHER KETONE (PEK) AND POLYETHER ETHER KETONE (PEEK)

Alternative names
Polyaryl ether ketones
Aromatic polyetherketones

Polymer structure

PEK

PEEK

Polymer status
High cost speciality engineering plastics. Normally brought into consideration when polysulphones do not meet specification.

Principal variants
In addition to the two basic polymers there has been particular interest in carbon fibre reinforced laminates produced by compression moulding techniques. Sheet for thermoforming is available at different levels of crystallinity.

Key application properties
High softening point, suitable for long term use above 200 °C. Low flammability (UL94 V–O rating at 1.5 mm thickness) and exceptionally low smoke emission. Excellent hydrolytic stability, very good stress cracking resistance and good radiation resistance.

Selected typical properties

	PEEK	PEK
Specific gravity	1.265–1.32	1.32
Tensile strength (MPa)	92	110
Modulus (MPa)	3900	
Elongation at break (%)	>40	
Impact strength (Izod notched ASTM D256)		
(ft lbf in^{-1})	1.52	1.52
kJ/m^2	8	8
Glass transition temperature °C	143	165
Crystalline melting point °C	334	365
Deflection temperature under load (1.82 MPa)	150	165

Coefficient of thermal expansion (1/°C $\times 10^{-5}$)	5.5
Volume resistivity (Ωcm)	10^{16}
Loss tangent 60 Hz	0.001
10 000 Hz	0.002
Permittivity 60 Hz–10 000 Hz	3.2

Chemical and solvent resistance The polyketones have overall excellent chemical resistance but may be attacked by concentrated mineral acids and by chlorine. Hydrolytic stability is outstanding with no loss in ductility after 2000 h at 260 °C.

Maximum service temperature PEEK can be used up to 240 °C with PEK somewhat higher.

Burning Properties Inherently low flammability meeting UL 94 V–O ratings. Toxic gas and smoke emission as low as any thermoplastics material.

Key processing properties
- PEEK can absorb up to 0.5% water in 24 h at 40% RH. Because of the high processing temperatures this must be reduced to below 0.15% before processing (e.g. by drying for 3 hours at 150 °C).
- Viscosity at typical moulding temperatures similar to that of a stiff flow ABS at typical processing temperatures for ABS.
- Melt has high level of stability at the high processing temperatures although there may be some change in melt viscosity. Generally recommended not to exceed processing residence time of 30 min.
- Shrinkage (for PEEK) c. 0.012 cm/cm for unfilled grades; filled grades 0.002–0.015 cm/cm according to fibre orientation.

Unlike some other engineering thermoplastics do not give off corrosive gases during processing so that corrosion resisting materials of construction are unnecessary.

Typical injection moulding conditions
The following data is for PEEK. Initial recommendations for PEK are for melt temperatures at 380–420 °C, barrel temperatures of 385–390 °C and mould temperatures of about 165 °C.

	Unfilled	Filled
Recommended melt temperatures (°C)	360–380	380–400
Temperature settings (°C)		
Barrel – (rear)	340–360	350–370
(middle)	360–370	370–380
(front)	370–380	380–390
Mould	160–170	160–175
Injection pressures	up to 150 MPa	

Shutting down and reprocessing Material should not be left in barrel on cooling because high adhesion between polymer and metal may cause damage. Purging with PE is recommended. Because of good thermal stability PEEK may be reprocessed using up to 30% reground material. PEK has been claimed to be more recyclable than PEEK.

Other comments: Because of high melt viscosity runners should be short and generous in size whilst large gates and tapered sprues are recommended for PEEK and PEK.

Typical extrusion conditions
PEEK and PEK may be processed on conventional extrusion equipment using melt temperatures in the range 350–420 °C. Since corrosive gases are not liberated during processing corrosion–resistant materials of construction are not necessary.

Thermoforming
Grades of PEEK are available varying in the level of crystallinity developed in the film during manufacture. Fully crystallized grades require forming temperatures of 340–350 °C. This will destroy any crystallization. For a high level of crystallinity in the formed product mould temperatures should be 180–240 °C; for a lower level of crystallinity moulds may be 25–130° C. Partially crystallized grades may use forming temperatures of 170–180 °C and mould temperatures of about 120 °C.

Applications
Used in critical and hostile environments. Of particular interest where heat resistance and hydrolytic stability important. Used for abrasion and chemically resistant linings, pump impellers, aircraft components. Carbon fibre laminates are used structurally in aircraft whilst thermoformed sheet may be used for aircraft interior trim.

POLYETHYLENE

Alternative names
Polythene
Polyethene
Polymethylene (this term is usually reserved for laboratory materials of high regularity of chemical structure).

Polymer structure

$$\sim CH_2 - CH_2 \sim$$

Polymer status
The most important plastics material in tonnage terms. Of considerable importance for film, blow moulding and injection moulding.

Principal variants
LDPE Low density polyethylene – original commercial material; irregular structure with short and long chain branching to give limited crystallinity.

HDPE High density polyethylene – much less branching and as a result higher level of crystallinity.

LLDPE Linear low density polyethylene – short chain branching only but limited crystallinity.

Also MDPE (medium density), and VLDPE (very low density).

Grades also vary in: molecular weight; molecular weight distribution; presence of second monomers (e.g. vinyl acetate as in EVA) additives.

Key application properties
Low cost. Easy processability. Excellent electrical insulation characteristics. Excellent chemical resistance. Toughness and flexibility. Reasonable clarity of thin films (LLDPE and LDPE).

Limitations include: low softening point, susceptibility to environmental stress cracking, susceptibility to oxidation, opacity in bulk, wax-like appearance, low hardness, rigidity and tensile strength. High gas permeability.

Selected typical properties

	LDPE	HDPE	EVA
Specific gravity	0.914–0.93	0.94–0.96	0.93–0.95
Tensile strength (MPa)	9–15	20–24	10
Elongation at break (%)	150–650	20–800	–
Impact strength	Tough	Tough	Tough
Glass transition temperature	Subject of debate. Transitions at both -120 and -20 °C		
Crystalline melting point °C	c.108	125–136	63–103
Deflection temperature under load			
(1.82 MPa)	Because material is so flexible		
(0.48 MPa)	this property has little meaning		
Vicat softening Point °C	77–98	116–136	59–81
Coefficient of thermal expansion			
cm/cm/°C $\times 10^{-5}$	8	12	–
Volume resistivity (Ωcm)	10^{17}–10^{20}	10^{16}–10^{19}	10^{14}–10^{17}
Power factor	0.0001–0.0002	*	0.014–0.0008

* Generally very low but will depend on extent of catalyst residues

Chemical and solvent resistance Excellent. No solvents at room temperature. Will dissolve in hydrocarbon and other solvents of similar solubility parameter as melting point is approached. Will however swell in solvents of similar solubility parameter. A number of agents cause environmental stress cracking including esters, hydrocarbons, silicones. Resistant to strong acids and bases. May be attacked by halogens. Subject to oxidation and UV attack but better than PP.

Maximum service temperature Very dependent on stress levels during service. As a guide 70 °C for LDPE and 100 °C for HDPE.

Burning properties Burns easily with a smoky flame.

Key processing properties
- Polymer does not absorb water although some additives may make compound hygroscopic.
- Flow properties very dependent on molecular weight and amount of branching but usual moulding grades may be described as easy flow.
- Polymer has good stability to processing provided oxygen is absent.
- Shrinkage is high (0.015–0.040 cm/cm) and depends on type, cooling rate and part thickness. May cause problems of sink marks, voids and warping.

Typical injection moulding conditions

	LDPE	HDPE
Recommended melt temperatures (°C)	180–260	200–260
Temperature settings °C		
Barrel – rear	120–200	160–200
middle	160–260	170–260
front	200–280	220–280
Nozzle	210–270	210–270
Mould	10–60	5–50
Injection pressures: MPa First stage up to	130	180
Second stage up to	80	150

Shutting down and reprocessing No need to purge with other material

Other comments: Difficult to mould to close tolerances

Typical extrusion conditions

	LDPE	HDPE
Temperature profile (°C)	120–160	170–220
Recommended screw L/D ratio	20–25/1	20–25/1
Recommended compression ratio	3/1–4/1	3/1–4/1

Other comments: Above data is for general extrusion. In certain cases conditions may be very different, e.g. for wire covering temperatures may be much higher.

Typical blow moulding conditions

Melt temperatures (°C)	165	210
Temperature profile °C	120–160	165–210
Mould temperature °C	23–26	23–26

Applications
LDPE main market is for film but also widely used for flexible injection mouldings and blow mouldings, piping, wire covering and rotational moulding.

HDPE widely used both for injection moulding and bottle blowing, increasing in use for wrapping film. Important applications in electrical and chemical industry.

LLDPE mainly of interest for film applications.

POLYETHYLENE TEREPHTHALATE

Alternative names
PET
PETP

Polymer structure

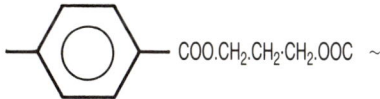

$\sim$ COO.CH$_2$.CH$_2$·CH$_2$.OOC $\sim$

Polymer status
Major polymer for synthetic fibres (e.g. Dacron, Terylene) and film (Melinex, Mylar). Limited use as an engineering thermoplastic, particularly when glass fibre filled. In recent years has captured large markets for bottles (particularly for beers and carbonated drinks).

Principal variants
Unfilled material may be processed to be largely amorphous or crystalline. Where high crystallinity required a *nucleating agent* to encourage crystal growth may be incorporated. Glass fibre filled and rubber-modified grades (e.g. Rynite Du Pont) are of major importance. Glass-filled grades are probably 90% of the crystalline PET market. About 70% of glass filled grades also contain flame retardant.

Key application properties
Crystalline grades similar to PBT but with higher modulus, gloss and heat distortion temperature.

Selected typical properties
Data given for grades with 30% glass fibre + fire retardant.

Specific gravity	1.6 (N.B. Unfilled amorphous 1.34 Crystalline 1.38)
Tensile strength (MPa)	130–150
Modulus MPa	10 100
Elongation at break %	2.0–2.3
Impact strength Izod J/m	77–90
Glass transition temperature °C	70–80
Crystalline melting point °C	250–255
Deflection temperature under load (1.82 MPa) °C	205–210 (N.B. Unfilled polymer 50 °C)

Volume resistivity Ωcm	10^{15}
Dielectric constant	3.5
Power factor	
1000 Hz	0.0055 (data for unfilled grade)
1 000 000 Hz	0.0208 (data for unfilled grades)

Chemical and solvent resistance Similar to PBT but as slightly more polar will be more resistant to hydrocarbons but slightly less resistant to strong acids and to absorption of water.

Maximum service temperature Usually about 120 °C but may be used at higher temperatures for short times.

Burning properties Unfilled grades will burn. Without flame retardant the limiting oxygen index is about 20 but is raised to about 29 by the use of flame retardants. (Such flame retarded glass filled grades will have UL 94 V–O ratings down to 0.7 mm).

Key processing properties
- Although moisture absorption is low (*c.* 0.2%) it is important to process dry material to avoid degradation and embrittlement. In bottle production moisture should be less than 0.005%. In latter case this may require drying in a dehumidifying drier for 4 h at 165 °C. For injection moulding 2 h in a dehumidifying drier or 4 h in an ordinary oven at 135 °C is normally satisfactory.
- Flow properties clearly depend on whether or not glass fibre is used. Unfilled materials may be rated as easy flow.
- PET is liable to degrade if overheated. It should not be heated above 300 °C and at 290 °C should not be held for more than 4 minutes. If material is not thoroughly dried degradation will be even more severe.
- Shrinkage is less with amorphous grades (0.0035 cm/cm in flow direction; 0.002 cm/cm in transverse) than with crystallizing grades (0.018 and 0.021 cm/cm). Filled grades exhibit much lower shrinkage, e.g. a 36% glass filled grade will have 0.002 cm/cm in flow direction but may be as high as 0.018 cm/cm in the transverse direction.

Use of nucleating agents can increase setting rate, reducing moulding cycles, and improve many mechanical properties.

Typical injection moulding conditions

Recommended barrel temperatures vary considerably between grades and this is reflected in the wide range given below. High mould temperatures encourage crystallization, particularly with nucleated grades giving fast setting whilst the low temperatures help to keep crystallinity to a minimum in the so-called amorphous grades.

Recommended melt temperatures (°C)	240–295
Temperature settings (°C):	
Barrel – rear	230–300
middle	240–300
front	250–300
Nozzle	250–300
Mould	80–120 (crystalline grades)
	15–25 (amorphous grades)

Injection pressures: First stage up to 160 MPa; second stage up to 130 MPa.

Shutting down and reprocessing Purge with HDPE or polystyrene. In the case of a stoppage purge out of barrel if stoppage less than 15 min; if more than 15 min purge with PE and reduce barrel temperatures to those required for PE until ready for restart.

Other comments: If reprocessing regranulate immediately and store in sealed containers. Ensure that stagnation does not occur because of under-utilization of barrel capacity.

Typical blow moulding conditions

Bottle grades are made by rapid quenching of melt into water in order to convert material to a clear amorphous state. The pellets are then further polymerized in the solid phase at temperatures just below the T_m. This solid phase process reduces the aldehyde content, which may impart a taste to bottled beverages and also reduces the quantity of extractable low molecular weight polymer.

Bottles are produced from amorphous pellet preforms by injection moulding into a cooled mould, reheating above the T_g, and then stretch blown in a cold mould to produce clear biaxially oriented bottles.

Thermoforming

While sheet is available in both amorphous and semi-crystalline forms it is the latter, sometimes known as C–PET, that is usually used for thermoforming. Its major advantage is that it is transparent to high frequency electromagnetic radiation and is used for food containers in microwave ovens, where its ability to be used in the range 175–205 °C is also of value.

Typical forming conditions (°C):	
Forming temperatures	150–175
Mould temperatures	150–175
Set temperature (i.e. temperature for	
removal from mould)	150

Applications

PET is best known for fibres, films and bottles. The market for engineering thermoplastics is much smaller and dominated by glass-filled grades. These materials are used largely on account of their stiffness and warp resistance. Uses include electrical/electronic components (e.g. electronic ignition rotors), non-electronic auto applications (e.g. window wiper components, rear view mirror housings), appliances (e.g. toaster, coffee maker and waffle iron components), medical devices, sporting goods and gear wheels.

POLYMETHYL METHACRYLATE

Alternative names
Acrylic (widely used but misleading since there are many acrylic materials including rubbers, surface coatings and fibres). PMMA. Poly[1–(methoxycarbonyl)–1–methyl ethylene] (This is the systematic name recommended by IUPAC).
Poly(methyl 2-methyl propenoate)

Polymer structure

$$\sim CH_2 - \underset{\underset{COO.CH_3}{|}}{\overset{\overset{CH_3}{|}}{C}} \sim$$

Polymer status
Major thermoforming material used particularly for baths and display signs. Special purpose injection moulding and extrusion material of particular value where high transparency and excellent weathering resistance are of importance.

Principal variants
Thermoforming material made by polymerization casting is of higher molecular weight than moulding and extrusion grades. Copolymers occasionally used. Fire retardant grades of some importance.

Key application properties
Noted for two outstanding properties – high light transmission of unfilled grades and excellent weathering resistance. Material is also hard, rigid and with a low moisture absorption. It is a good electrical insulator at low frequencies. There are a wide range of solvents and unmodified grades will burn in air.

Selected typical properties

Specific gravity	1.18–1.19
Tensile strength (MPa)	72
Modulus (MPa)	2400–3000
Elongation at break (%)	5
Impact strength (ft lbf in^{-1}) Izod (BS2782)	0.4
Glass transition temperature °C	104
Crystalline melting point	Polymer is amorphous
Deflection temperature under load °C	
(1.82 MPa)	85–100
(0.48 MPa)	88–116
Coefficient of thermal expansion cm/cm/°C	7×10^{-5}
Volume resistivity Ωcm	10^{14}
Dielectric constant	3.0–3.1

Chemical and solvent resistance Attacked by mineral acids but resistant to alkalis, water and most aqueous inorganic salt solutions. Soluble in a number of solvents with similar solubility parameters (e.g. carbon tetrachloride, chloroform, acetone and similar ketones). May crack or craze in a number of additional materials without dissolving (e.g. aliphatic alcohols).

Maximum service temperature 60–90 °C

Burning properties Grades without fire retardant burn with a blue flame giving off little smoke, but flaming melt liable to drip.

Key processing properties
- PMMA granules tend to pick up moisture (up to 0.3%) and care should be taken in storage and when using reground materials.
- Melt has high melt viscosity and processing equipment needs to be rugged. Flow path ratios vary from 100:1 to 150:1 according to grade used.
- Polymer has limited thermal stability and may depolymerize during prolonged exposure to high processing temperatures. Screws with decompression zones may be used to help remove volatile monomer produced during processing (monomer boils at 100 °C and may cause bubbles and mica marks similar to those produced by moisture). Alternatively predry 4 h at 70–75 ° C.

Typical injection moulding conditions

Recommended melt temperatures (°C)	225–240
Recommended temperature settings (°C)	
Cylinder – rear	150–180
middle	180–230
front	220–240
Nozzle	225–245
Mould	60–80
Injection pressures (MPa):	
1st stage	up to 175
2nd stage	up to 105

Shutting down and reprocessing High melt viscosity makes purging very difficult. Where necessary may have to strip down and clean.

Other comments: Optical imperfections may often be reduced by use of tab gates. Generous mould venting recommended to facilitate removal of volatiles. Good temperature control essential. Avoid overworking in barrel.

Typical extrusion conditions

Temperature settings 190–230 °C

Recommended screw L/D ratio 27:1 to 33:1

Recommended compression ratio 2.5–4.0

Other comments: Where extruded for use in optical applications take care to avoid marking and distortion on cooling. Air cooling common. Polymer usually rubbery on emergence from die which enables some post-extrusion shaping before cooling. Use light filter pack to avoid excess working of melt leading to degradation. For similar reasons use low screw rotational speeds and use screw with deep cut flights to minimize shear.

Thermoforming

For most applications thickness of sheet used and high modulus in the rubbery state preclude use of simple vacuum forming. Systems using compressed air and mechanical shaping (often in conjunction with vacuum) commonly used. Typical forming temperatures 150–190 °C.

Applications

Uses arise largely from high optical clarity and excellent weathering resistance. Injection mouldings widely used for car lamp covers and dashboard fittings. Extrudates of importance for street light fittings. Of interest as an optical fibre. Sheet material widely used for baths and other sanitary fittings, for signs and for lighting. Of use in dentures, spare part surgery, hygroscopic copolymers used in soft contact lenses.

POLYPROPYLENE

Alternative names
Polypropene (Also commonly referred to as propylene, which is incorrect, and by the slang term polyprop(e)).

Polymer structure

$$\sim CH_2 - CH \sim$$
$$|$$
$$CH_3$$

Polymer status
In tonnage terms the number three plastics material after PVC and polyethylene. Particularly important for injection moulding where it is commonly the first material to be considered for a new application.

Principal variants
The polymer molecule can be arranged in a variety of ways of which the *isotactic* form, a regular crystallizable structure, is the basis of most moulding and extrusion materials. An irregular amorphous form known as *atactic* polypropylene has a number of uses for example in bituminous compositions and for carpet backing.

The substantially isotactic materials vary in: presence of comonomers (many grades are based on a polymerization feed of propylene and ethylene); molecular weight; level of isotacticity; presence of glass fibre; presence of other additives particularly mineral fillers and blowing agents.

Key application properties
Compared with PE has: a lower specific gravity (*c.* 0.90); higher softening point (can withstand boiling water and be subject to many steam sterilizing operations); virtual freedom from stress cracking; higher brittle point; is more susceptible to oxidation.

Like PE has very low water absorption, excellent chemical and solvent resistance and very good electrical insulation characteristics. Moulding surfaces tend to be harder, of higher gloss and more mar resistant than those of PE but not generally as good as ABS in this respect.

Selected typical properties
The data given below are for unfilled grades; where a broad spread of values is quoted this indicates that the property is sensitive to the grade of material used:

Specific gravity	0.90
Tensile strength (MPa)	25–35
Flexural modulus MPa	100–1300
Elongation at break %	35–350
Impact strength	Wide range from brittle to tough
Glass transition temperature (°C)	Homopolymer c. 0 °C; Copolymers may be as low as −60 °C
Crystalline melting point °C	145–150
Deflection temperature under load	This is somewhat meaningless since the flexibility of the material will give low values that do not reflect the high service temperature possible with PP.
(1.82 MPa) °C	53–57
(0.48 MPa) °C	96–102
Coefficient of thermal expansion cm/cm/°C	8×10^{-5}
Volume resistivity Ωcm	10^{16}–10^{17}
Power factor	0.0004–0.001

Chemical and solvent resistance Excellent chemical resistance. There are no solvents at room temperature, very few materials that cause environmental cracking and very few that chemically attack it (e.g. halogens). It is however subject to oxidation, particularly at elevated temperatures, on exposure to light and in contact with copper.

Maximum service temperature Depends on level of loading; in the range 70–135°C.

Burning properties Burns with a blue flame

Key processing properties
- Does not absorb water.
- Flow properties depend on molecular weight and additives present. Unfilled grades generally considered as easy flow although temperatures used are somewhat higher than for PE. Flow path: wall thickness ratios of 175:1 are possible on 1 mm wall thickness sections. With easy flow grades the ratio may be as high as 350:1.
- Thermal stability is quite good in the absence of oxygen so that there is no need to purge with another material when shutting down.

- Because of its crystallinity there is a high moulding shrinkage, typically of the order of 0.020 cm/cm and is reasonably uniform in all directions. As with many crystalline polymers shrinkage tends to be higher with thicker sections because of the longer cooling times.

Typical injection moulding conditions

Recommended melt temperatures (°C)	230–275
Temperature settings °C	
Barrel – rear	170–210
middle	210–250
front	230–260
Nozzle	240–270
Mould	5–80

Injection pressures Machine should be capable of giving up to 180 MPa first stage and 150 MPa second stage.

Shutting down and reprocessing No need to purge when shutting down. Up to 20% reground material may be used with virgin materials for less critical applications.

Other comments: It may be necessary to apply high dwell pressures for quite lengthy periods to avoid excessive shrinkage, sink marks and voids. Avoid overpacking.

Typical extrusion conditions

Melt temperature °C	240
Temperature profile along barrel °C	160–240
Recommended screw L/D ratio	18:1 to 25:1
Recommended compression ratio	3:1 to 4:1

Other comments

With thick sections use zoned cooling to reduce voids. Avoid copper and brass tools in order to prevent copper-related oxidation.

Typical blow moulding conditions

Extruder type and temperature profile as under typical extrusion conditions. Dies should be about 20–25 deg C below maximum barrel temperature. Moulds cold (20–25 °C) or even refrigerated for high output.

Thermoforming

Although widely used is not easy to thermoform because of the narrow thermoforming range. Hot melt strength is poor at forming temperatures and care must be taken to avoid excessive sag. For sheet > 2 mm thick, double sided heating is necessary.

Temperature settings (°C)	
Forming temperatures	140–170
Set temperature*	85

* Temperature at which forming may be removed from mould without undue distortion.

Applications

Very widely used on account of its low cost, low water absorption, excellent chemical resistance, good appearance, good electrical insulation properties and good if not outstanding heat resistance. Widely used in household goods, car components, toys, medical equipment, luggage, integral hinges. Used to some extent for housings although ABS (q.v.) generally preferred on account of potentially better surface finish and mar resistance. Also of importance in film and fibre form.

MODIFIED PPO

This section gives details on the General Electric product *Noryl*. This is a blend of poly–2,6–dimethyl–*p*–phenylene oxide (often incorrectly referred to as polyphenylene oxide or PPO) with polystyrene or some related material. In recent years a number of similar proprietary materials have come onto the market. This material should not be confused with polypropylene oxide used mainly in the manufacture of polyurethane foams which is also sometimes referred to by the abbreviation PPO.

Alternative names
See note above.

Polymer structure

~ ═ o ~ with polystyrene (*q.v.*)

Polymer status
Major engineering thermoplastic.

Principal variants
Polymer ratio in blend; type of styrenic blending polymer used; self-extinguishing grades; glass-reinforced grades; structural foam grades.

Key application properties
Ability to work to close dimensional tolerances as a result of low moulding shrinkage, low coefficient of expansion and low water absorption. Excellent resistance to hydrolysis. Very good electrical properties over a wide range of temperature. High heat distortion temperature. Good flame resistance of some grades.

Selected typical properties

	Unfilled grades	Glass fibre filled grades
Specific gravity	1.06	1.21–1.46
Tensile strength (MPa)	48–76	65–124
Modulus (MPa)	2200–2500	2900–7600
Impact strength (Izod) (ft lbf in^{-1})	5–10	2–4
Glass transition temperature	The blend may show two transitions (At 116 and 150 °C)	
Crystalline melting point	257 °C (for PPO component)	
Deflection temperature under load (1.82 MPa)	88–149 °C	88–150 °C
Coefficient of thermal expansion 10^{-5} m/m/°C	6.5–8.3	2.5–5.4
Dissipation factor 60 Hz @ 23 °C	0.0004–0.005	0.0008–0.005

Chemical and solvent resistance Good resistance to acids and bases. Dissolved by some aromatic and halogenated solvents such as trichloroethylene, chloroform, xylene and toluene. A number of materials will cause cracking and crazing.

Maximum service temperature Underwriters Laboratories give values ranging from 50 to 110 °C according to the grade used and the test method.

Burning properties In the absence of flame retardants these materials are given UL classifications of HB and V–1. Modified grades may have V–O and 5V ratings. One grade has a 5V rating down to 20 mm thickness.

Key processing properties
- These materials have very low water absorption for an engineering thermoplastic (typically 0.07% in 24 h at 23 °C) so that predrying is not normally necessary. Where it is necessary then predry for about 2 h at 100 °C (130 °C for glass-filled grades).
- Flow properties depend on grade but are similar to those for ABS. A typical melt temperature is about 145 deg C above the heat distortion temperature.
- The material will degrade on overheating; too long a drying time may cause similar faults (e.g. splay marks) to those encountered with damp material.
- Moulding shrinkage is typical for that of an amorphous material (0.005–0.007 cm/cm); in the case of 30% glass-filled grades shrinkage may be as low as 0.002 cm/cm.

Shrinkage values are similar along and across the flow direction.

Typical injection moulding conditions

	Unfilled	Filled
Recommended melt temperatures (°C)	260–300	290–325
Temperature settings °C:		
Barrel – (rear)	210–250	240–260
(middle)	230–270	270–290
(front)	250–290	280–310
Nozzle	240–275	265–295
Mould	60–110	90–110

Injection pressures: First stage up to 180 MPa; second stage 140 MPa. Second stage pressure critical in controlling sink marks and residual stresses.

Shutting down and reprocessing Purge with PS or HDPE

Other comments: Take care to avoid 'dead' spots where melt can stagnate; avoid shut-off nozzles and mixing heads. Low cycle times are possible because of rapid setting of polymer in mould cavity.

Typical extrusion conditions

Temperature settings: Depend on grade used. Typically the melt temperature at the die should be (H+100)°C where H is the heat distortion temperature (deflection temperature) of the grade being extruded.
Recommended screw L/D ratio: Minimum 24/1
Recommended compression ratio: 2.2/1 to 3.5/1

Other comments: Air cooling is preferred to minimize stresses and warping in extrudate.

Applications
Low shrinkage, low water absorption and low coefficient of expansion enable close tolerances to be operated. This is useful in components used in contact with water (e.g. washing machines, dishwashers, water pumps). Good heat resistance, impact strength and flame retardance lead to many applications in electronics and the automotive industry. Structural foam grades of importance for housings.

POLYPHENYLENE SULPHIDE

Alternative names
PPS

Polymer structure

Polymer status
Special purpose engineering thermoplastic with high levels of heat resistance.

Principal variants
Because of the low T_g and the tendency to brittleness, commercial grades are normally filled. There are three main types: glass fibre filled, e.g. 40% glass fibre; glass fibre/mineral filled; carbon fibre filled, e.g. at 30% level of fibre.

Higher molecular weight grades are less brittle.

Key application properties
Excellent heat resistance for a thermoplastic with UL temperature indices of about 240 °C. Filled grades have high deflection temperatures. Member of select group of such heat resisting injection mouldable materials including the polysulphones, polyetherimides, polyether ketones and polyether ether ketones. Also shows excellent flame retardancy without use of additives, very good chemical resistance and very good electrical insulation properties. Is also attractive because of its precision mouldability and high dimensional stability (low moulding shrinkage and low water absorption).

Selected typical properties

	40% GF filled	Glass/ mineral filled	30% carbon fibre filled
Specific gravity	1.6	1.8	1.45
Tensile strength MPa	135	92	186
Flexural modulus MPa	11,700	13,100	16,900
Elongation at break %	1.3	0.7	–
Impact strength Izod notched J/m	75	27	58

Glass transition temperature °C	<--- 85 °C for unfilled material --->		
Crystalline melting point °C	<--- 285 °C for unfilled material --->		
Deflection temperature under load (1.82 MPa)	<----> 260 °C for each type ----> (NB If low mould temperatures are used the deflection temperature may be as low as 150 °C due to lower crystallinity – see below)		
Volume resistivity Ωcm	4.5×10^{16}	2×10^{15}	1.3×10^{15}
Dissipation factor 1 MHz	0.0013	0.016	0.019

Chemical and solvent resistance Is attacked by concentrated mineral acids, some amines and benzaldehyde. Some slight/moderate attack (either chemical change or swelling) by a variety of reagents including sodium hydroxide, sodium hypochlorite, methyl ethyl ketone, chloroform, aniline, toluene and phenol. Exposure to water, alcohols and air may cause a slight drop in tensile strength over a one year period. Only moderate stability to UV light unless stabilized. May be cross-linked by air oxidation at elevated temperatures.

Maximum service temperature UL temperature indices range from 200 to 240 °C according to test method used and to grade. Amongst thermoplastics only the polyketones amongst established materials achieve such high figures.

Burning properties When exposed to an external source of flame it will continue to burn with a yellow-orange flame until the external flame source is removed. It does not drip and the char is black and glossy. Filled grades usually have limiting oxygen index values in excess of 46 and have UL 94 V–O rating.

Key processing properties
• Water absorption of the polymer is very low (c. 0.05%) but filled grades exhibit somewhat higher values (presumably due to capillary action along fibre–polymer interface). All material should therefore be dried before moulding or extrusion, typically in a desiccator drier for 3–6 h at 150–175 °C.

- At the high melt temperatures used (300–360 °C) the melts may be rated as easy flow.
- Although the materials have good thermal stability, should not be exposed to temperatures above 375 °C during processing, to reduce evolution of irritating gases. In presence of air cross-linking may occur at high temperatures.
- The filled (i.e. commercial) grades exhibit a low moulding shrinkage (0.001–0.005 cm/cm) usually being twice as great transverse to the flow direction than in the flow direction. Thicker sections may shrink more than thin so that in some mouldings warping may occur. In injection moulding the mould temperature is also significant. If the mould is well below the glass transition temperature (c. 85 °C) the moulding will show a low level of crystallinity although this may be increased by a post-moulding annealing process at temperatures above the T_g. Higher mould temperatures (120–140 °C) will require longer cycle times but with higher crystallinity, increased dimensional stability, increased deflection temperature and a smooth glossy surface.

Typical injection moulding conditions

Recommended melt temperatures (°C)	300–360
Temperature settings (°C):	
Barrel – rear	290–310
middle	300–330
front	315–360
Nozzle	305–340
Mould	30–80 for low crystallinity
	120–160 for high crystallinity
Injection pressures:	Machine should be capable of up to 150 MPa for the first stage and 100 MPa for the second stage. Use highest pressure that does not cause flashing.

Shutting down and reprocessing For overnight stops empty barrel, leave screw forward and turn off heaters. For cleaning purge with an extrusion grade of HDPE.

Other comments: Moulds need to be properly vented with vents no more than 0.0075 mm deep and 6.35 mm wide on the parting line. Drooling may be a problem and the use of shut-off valves is generally to be preferred.

Compression moulding
There is some use for compression moulding of PPS; mainly to produce cross-linked material. In a typical process the raw polymer is blended with fillers and then precured in a shallow pan in air for 12–16 h at c. 270 °C (i.e. below the T_m). It is then given a further cure for 1.5–2 h at 325–360 °C and the material then cooled and granulated for moulding. The granules are cold pressed at 12–18 MPa and then heated in the mould until a minimum temperature in the polymer mass of 305 °C is achieved (typically this will require 1–3 h at 345–370 °C). Controlled cooling is necessary to avoid cracking and voids.

Applications
Connectors, coil formers, bobbins, relay equipment; for metal replacement in the automotive industry (carburettor parts, ignition plates, lamp sockets); street lamp reflectors; pH meter components. Encapsulation compounds.

POLYSTYRENE

Alternative names
Polyphenyl ethene
Polyphenyl ethylene

Polymer structure

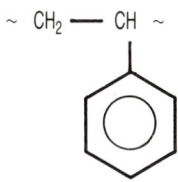

~ CH$_2$ — CH ~

Polymer status
Major, general purpose, low cost thermoplastic

Principal variants
Variants with unmodified polystyrene include variations in molecular weight. Commercial materials are essentially atactic but more regular polymers have been prepared. Important modifications include high impact polstyrene (HIPS), ABS (q.v.), *PPO/PS blends* (q.v.) and expanded polystyrenes. Only unmodified (GPPS) and HIPS are dealt with in this section. In recent years there has been a trend to use the term *toughened polystyrene,* with abbreviation TPS, instead of high-impact polystyrene.

Key application properties
Rigidity. Transparency. Low moisture absorption. Excellent electrical insulation characteristics. Low density. Low raw material cost.

Disadvantages include wide range of solvents, brittleness of GPPS, comparatively low maximum service temperature, and ability to burn readily with a smoky flame.

Selected typical properties

	GPPS	HIPS
Specific gravity	1.054	1.02–1.05
Tensile strength (MPa)	40–56	14–40
Modulus MPa	2750–3750	1750–3250
Elongation at break (%)	1.0–2.5	5–20
Impact strength ft.lb/in notch (Izod)	0.25–0.35	0.6–5.0
Glass transition temperature °C	90–100	

Crystalline melting point	Amorphous – not applicable	
Deflection temperature under load °C		
(1.8 MPa)	80	75–80
(0.48 MPa)	91	80–90
Coefficient of thermal expansion cm/cm/°C	3×10^5	
Volume resistivity Ωcm	10^{16-18}	
Power factor (60–10^6 Hz)	0.0002–0.0004	

Chemical and solvent resistance Dissolved by many hydrocarbons such as benzene, toluene and ethyl benzene, by many chlorinated hydrocarbons, by some ketones but not acetone, some esters and oils. May be crazed by many acids, alcohols and oils that do not act as solvents, e.g. white spirit.

Key processing properties
- Unmodified grades have negligible water absorption.
- Melts are of 'medium' viscosity but highly pseudoplastic. Typical flow path: wall thickness ratio is 150:1 for GPPS and 130:1 for HIPS. Easier flow HIPS grades tend to have lower softening points. The low specific heat and absence of a latent heat result in a low heat requirement for melting.
- Melts have good stability at processing temperatures.
- Low moulding shrinkage c. 0.005 cm/cm is typical of an amorphous polymer.

Typical injection moulding conditions

	GPPS	HIPS
Recommended melt temperatures (°C)	200–250	180–250
Temperature settings (°C):		
Barrel – rear	150–180	160–190
middle	180–230	170–230
front	210–280	190–250
Nozzle	210–280	180–240
Mould	10–80	10–80

Injection pressures: Low. The machine should be capable of giving
1st Stage   150 MPa
2nd Stage   75 MPa

Shutting down and reprocessing Polymer stability is quite good and it is not normally necessary to purge when shutting down. Stripping down and cleaning is advisable if changing to another material. Recycling percentage should not exceed 15–20%.

Other comments: Packing pressures should be kept low and follow up times short to avoid excessive internal stresses.

Typical extrusion conditions

Temperature profile 150–220 °C

Recommended screw L/D ratio 25:1 to 30:1

Recommended compression ratio 2:1 to 3:1

Other comments: General purpose polystyrene is not a common extrusion material other than as an intermediate stage in making biaxially oriented sheet and bottle manufacture. HIPS is widely extruded to make sheet used in thermoforming.

Blow moulding
Polystyrene is not commonly extrusion blow moulded but rather injection blow moulded, the mouldings being of particular interest in cosmetics and pharmaceuticals.

Thermoforming
HIPS is one of the major thermoforming materials. Biaxially stretched polystyrene is also widely used in blister packaging.
  Forming temperatures 130–180 °C
  Typical setting temperatures 85 °C (temperature at which forming may be removed from mould)

Applications
GPPS is used for toys, containers, tape cassettes, disposable low cost tumblers, display products. HIPS is widely used for appliance housings, furniture components, egg boxes, dairy product containers, refrigerator liners. Expanded polystyrene is used for insulation and for packaging. When used in conjunction with food it is important that levels of styrene monomer and of processing degradation products are acceptably low.

POLYSULPHONE

Alternative names

Polysulfone. Some types are referred to as polyethersulphones and others as polyarylsulphones. Although this is somewhat misleading as all commercial materials could be considered as polyethersulphones or as polyarylsulphones, this section will follow standard practice and distinguish between polysulphones (PSU) and polyethersulphones (PES).

Polymer structure

Polymer status

Specialist engineering thermoplastic, usually only considered after it has been established that polycarbonates do not meet the specification. In turn polyethersulphones usually only considered after it has been established that the cheaper polysulphone does not meet requirements.

Principal variants

Besides the basic polysulphone and polyethersulphone there are specialized grades with even higher levels of heat resistance. Glass fibre and carbon fibre filled grades are also available.

Key application properties

High temperature resistance with both a high softening point and resistance to chemical change on heating. UL continuous use temperatures may be 180 °C. Exceptional resistance to creep for an unfilled material, rigidity, transparency and self-extinguishing. Smoke evolution of polyethersulphones is extremely low. As with polycarbonates tough but notch sensitive. Good electrical insulation, but as with most aromatic polymers tracking resistance is limited.

Selected typical properties

	Unfilled	PSU 40% G.F.	PES
Specific gravity	1.24	1.54	1.37
Tensile strength (MPa)	70	124	84
Tensile modulus (MPa)	2600	11000	–
Flexural modulus (MPa)	2100	–	2570
Elongation at break (%)	50–100		
Impact strength (ft lbf in^{-1}) notch	1.3	–	1.6
Glass transition temperature (°C)	190	–	230
Crystalline melting point (°C)		Polymer is amorphous	
Deflection temperature under load (°C)			
(1.82 MPa)	174	187	203
Coefficient of thermal expansion (cm/cm/°C)	5.6×10^{-5}	2×10^{-5}	5.5×10^{-5}
Volume resistivity Ωcm	10^{16}	–	10^{17}–10^{18}
Power factor 60 Hz	0.001	–	0.001
1 MHz	0.006	–	0.006

Chemical and solvent resistance Generally resistant to aqueous acids and alkalis but attacked by concentrated sulphuric acid. Resistant to aliphatic hydrocarbons. Solvents include dimethyl formamide and dimethyl acetamide. Other liquids may cause cracking (e.g. ethyl acetate and acetone and, to a lesser extent, carbon tetrachloride.)

Maximum service temperature (°C) 160 (PSU), 180 (PES) (UL Ratings)

Burning properties Self-extinguishing. PSU has an LOI of 30, PES 33–38. Grades of both types have a UL V–O rating.

Key processing properties

- These polymers do tend to absorb water, up to 2%, and at the high processing temperatures this can be troublesome by leading to streaks and splash marks. Moisture does not however cause hydrolysis. Material should be dried 3 h at 150 °C or 4 h at 135 °C before moulding.
- Melts have high viscosity even at the high processing temperatures. Flow path: wall thickness ratios are of the order of 60:1 to 120:1 for unfilled grades and as low as 30:1 for glass-filled grades. Processing equipment needs to be robust.
- Polymer will darken and degrade on overheating. With time a black layer of degraded polymer may be built up on

49

barrel/cylinder walls and then pull away giving black marks on the moulding or extrudate. This necessitates periodic purging.

- Low shrinkage is typical of amorphous materials (0.006–0.007 cm/cm in both flow and transverse directions. May be as low as 0.002–0.003 cm/cm with glass-filled grades.)

Typical injection moulding conditions

	PSU	PES
Recommended melt temperatures (°C)	350–380	
Temperature settings °C:		
Barrel –rear	295–355	335–355
middle	310–370	345–365
front	315–380	350–370
Nozzle	320–380	355–375
Mould	100–150	140–160
Injection pressures:		
1st Stage	up to 200 MPa	
2nd Stage	up to 120 MPa	

Shutting down and reprocessing Turn off heat and purge until clear. If restarting with polysulphone maintain temperature at 200 °C until start up. If changing to another material purge with a high melt viscosity polyethylene or polypropylene.

Other comments: The level of frozen-in stresses may be gauged by the tendency to crack in ethyl acetate. In highly stressed mouldings the cracking time may be as little as 2 s. Generous dimensions for runners and sprues are recommended.

Typical extrusion conditions

Temperature profile 340–390 °C
Recommended screw L/D ratio > 20:1
Recommended compression ratio 2.5:1

Applications
Sterilizable medical equipment. Components of hot water installations. Electronic components including injection moulded printed circuit boards, moulded complete with holes, stand-offs and connectors. Aircraft components. Appliances requiring very good heat resistance.

POLYTETRAFLUOROETHYLENE (PTFE) AND TETRAFLUOROETHYLENE–HEXAFLUORO-PROPYLENE COPOLYMER (FEP)

Alternative names

Polytetrafluoroethylene is most commonly known by its abbreviation PTFE. The IUPAC systematic name is poly(tetrafluoroethene).

The copolymer FEP is also sometimes known (incorrectly) as fluorinated ethylene-propylene and sometimes as FEP fluorocarbon.

Polymer structure

Polymer status

PTFE is an important speciality plastics material used in a wide variety of applications. FEP is a less commonly encountered injection mouldable copolymer.

Principal variants

PTFE is available as a granular polymer, used for shaping dry granules; as a dispersion polymer which is shaped in paste form and as a latex. FEP is available as a conventional injection moulding material.

Key application properties

PTFE: Excellent heat resistance. Excellent electrical insulation properties over a wide range of temperature and frequency (probably not excelled by any other plastics material). Excellent chemical resistance. Negligible water absorption. No solvents. Very low coefficient of friction to metal. Excellent non-stick properties. Tough. Disadvantages include low stiffness, considerable difficulty in processing and high raw material cost.

FEP: This is easier to process than PTFE and may be injection moulded. Softening point is slightly lower but the polymer otherwise has similar properties to PTFE.

Selected typical properties

	PTFE	FEP
Specific gravity	2.1–2.3	2.16
Tensile strength (MPa)	17–21	19–22
Modulus (MPa)	500	
Elongation at break (%)	200–300	250–350
Impact strength ASTM Izod (ft lbf in^{-1})	2.0	2.9
Glass transition temperature °C	115	–
Crystalline melting point °C	327	270–290
Deflection temperature under load (0.48 MPa) °C	121	88
Coefficient of thermal expansion (μk^{-1})	100	
Volume resistivity Ωcm	> 10^{20}	> 10^{20}
Power factor	< 0.0003	< 0.0003

Chemical and solvent resistance The chemical resistance of PTFE and FEP is exceptional. There are no solvents and they are attacked at room temperature only by molten alkali metals and in some cases by fluorine. Treatment with a solution of sodium metal in liquid ammonia will sufficiently alter the surface of a PTFE sample to enable it to be cemented to other materials using epoxide resin adhesives. Weathering resistance is outstanding but they are degraded by high energy radiation.

Maximum service temperature For short term (< 200 h) 230–300 °C: for long term (1000 days) 150–220 °C.

Burning properties Polymers are very difficult to ignite and PTFE has an LOI of 90. Decomposition products may however be toxic.

Key processing properties

- Neither PTFE nor FEP absorb water so that drying is not normally necessary unless the material has been allowed to get wet.
- PTFE is too viscous (c. 10^{11} poise at 350 °C) to be processed on conventional thermoplastics processing equipment and special techniques are required for shaping.

FEP may be shaped by injection moulding but it is very viscous and melt fracture occurs at low shear rates (c. 100 s^{-1}).

- FEP may degrade if overheated during moulding operations and this may be accompanied by internal delamination.
- The amount of shrinkage occurring with PTFE depends on the process used. Moulding shrinkage with FEP is high (0.030–0.060 cm/cm).

Typical injection moulding conditions
The following data is for FEP.

Recommended melt temperatures (°C):	300–380 (typically 350)
Temperature settings (°C)	
Barrel – rear	315–330
middle	330–350
front	345–370
Nozzle	345–370
Mould	200–240

Injection pressures: These are kept to a minimum to reduce the probability of delamination in the gate area. The machine should be capable of giving

1st Stage	up to 105 MPa
2nd Stage	up to 52 MPa

Shutting down and reprocessing Reduce all settings to 315 °C and slowly purge until barrel is empty. Leave screw in forward position. To remove FEP from system purge dry as above then reduce temperature further to 290 °C and purge with HDPE.

Other comments: Degradation products may be quite unpleasant and exhaust hoods should be fitted close to the nozzle, mould and cooling areas. These products are also corrosive and screws and barrels should be corrosion resistant. Use of back pressure during mould filling of up to 60 MPa may be useful in evenly raising melt temperature.

Screw smear heads are preferable to non-return valves, to reduce chances of stagnation and hence degradation and delamination.

Typical extrusion conditions

The following data is for FEP.

Temperature settings 300–380 °C
Recommended screw L/D ratio 20:1
Recommended compression ratio 3:1

Other comments: Nylon-type screws have been successfully used.

Other processes
The processes used for PTFE are derived from those used in powder metallurgy. They generally involve a shaping operation compressing a powder or deformation of a PTFE paste. The shaped article is then heated to a high temperature to sinter it, sometimes under pressure.

Applications
PTFE is widely used because of its chemical inertness, exceptional weathering resistance, excellent electrical insulation properties, excellent heat resistance, very low coefficient of friction and non-adhesive properties. Chemical uses include seals, gaskets, packings, pump parts and laboratory equipment. Electrical uses include printed circuit boards, wire insulation, hermetic seals for condensers. Other uses include chute linings, flexible steam hose, unsintered tape for pipe sealing, aircraft components and non-stick ovenware. FEP is used for similar applications in the chemical and electrical industries.

POLYVINYL CHLORIDE

Alternative names
PVC
Poly-1-chloroethylene
Polychloroethene
Also commonly, but misleadingly, referred to simply as vinyl.

Polymer structure

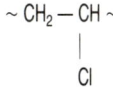

Polymer status
In tonnage terms second only to polyethylene.

Principal variants
There are two main types
UPVC unplasticized PVC – a rigid material.
PPVC plasticized PVC – contains plasticizer (from 15–100 pts per hundred parts polymer) and will be flexible or rubbery.
Copolymers, e.g. with vinyl acetate have some uses, whilst UPVC may be blended with rubbery or other polymers to produce toughened blends.

Key application properties
UPVC: Rigidity. Hydrocarbon resistance. Good weathering resistance. Good chemical resistance. Low cost. Low flammability.
PPVC: Flexible. Otherwise similar to UPVC above although plasticizer may reduce chemical resistance, weatherability and flame retardancy. (Note that noxious gases are released when the material does burn.) Good electrical insulation properties at low frequencies.

Selected typical properties
The properties of PVC compounds depend considerably on formulation. The figures below must be treated with considerable reserve.

	UPVC	PPVC
Specific gravity	1.4	1.3
Tensile strength (MPa)	58	10–30
Modulus MPa	2800	3–20
Elongation at break %	5	100–450

Impact strength (ft lbf in^{-1}) notch	1–18	tough
Glass transition temperature °C	c. 80 °C	–
Crystalline melting point	None – polymer is amorphous	
Deflection temperature under load °C (1.82 MPa)	76–80	–
Coefficient of thermal expansion cm/cm/°C	5×10^{-5}	
Volume resistivity Ωcm	$> 10^{14}$	10^{12}–10^{15}
Power factor	0.02	0.08–0.10

Chemical and solvent resistance Very good chemical resistance. Resistant to hydrocarbon solvents. Low water absorption. Dissolved by cyclohexanone, dichloroethane and nitrobenzene.

Maximum service temperature °C
70 for UPVC and 60–105 for PPVC.

Burning properties Self-extinguishing but PPVC properties will depend on plasticizer used.

Key processing properties
- Polymer does not absorb water but some plasticizers may do so.
- UPVC melts are viscous. Typical flow path ratio c. 60:1. Viscosity of PPVC depends on plasticizer level.
- Polymer lacks thermal stability. Degradation rapid at elevated temperatures to avoid dead spots in machines. Ram injection moulding machines are not satisfactory. *Never allow polymer to mix with polyacetals since combinations at elevated temperature may lead to explosions.*
  Because hydrochloric acid (HCl) is given off due to degradation during processing all metal surfaces that come in contact with the melt should be resistant to this acid. Other metal surfaces should also be protected. Good ventilation of the working area is also essential.
- Polymer is amorphous so that shrinkage is low c. 0.005 cm/cm.

Typical injection moulding conditions

	UPVC	PPVC
Recommended melt temperature (°C)	180–200	150–190
Temperature settings °C:		
Barrel – rear	140–160	140–155
middle	160–180	155–175
front	180–190	175–180

| Nozzle | 190–205 | 180–185 |
| Mould | 20–60 | 20–30 |

Injection pressures (MPa):		
First stage up to	175	120
Second stage up to	100	80

Shutting down and reprocessing On completion of run turn off heaters and purge slowly. Special purging compounds of enhanced stability are available; alternatively use PS. Never purge with acetals.

Other comments: All metal surfaces that come in contact with PVC melts should be resistant to hydrochloric acid. Very good temperature control is necessary. Never allow material to hang up in the barrel.

Typical extrusion conditions

	UPVC	PPVC
Temperature profile (°C)	150–180	140–175
Recommended screw L/D ratio	14:1–17:1	17:1–20:1
Recommended compression ratio	c. 2:1	c. 2:1

Other comments: With UPVC it is difficult to pump granules in the feed zone in an efficient manner and for this reason twin-screw extruders are sometimes preferred. With both UPVC and PPVC a single start steady taper screw is advisable when using a single screw extruder.

Typical blow moulding conditions
Blow moulding is confined to UPVC using general conditions given above under extrusion conditions. Beryllium copper alloy moulds combine corrosion resistance with the ability to take and keep highly polished surfaced finish.

Other processes
Calendering is an important process for making PVC sheet. Plasticized PVC may also be processed in paste form (a physical mixture of polymer, plasticizer and other ingredients), which may be shaped by such techniques as rotational moulding and spreading, and then gelled by heating at elevated temperatures to make tough homogeneous products. PVC may also be thermoformed. While UPVC may be used it is more common to use modified materials such as vinyl chloride–vinyl acetate copolymer, chlorinated PVC (CPVC), UPVC containing an acrylic impact modifier and ABS/PVC blends.

Applications
UPVC is widely used in building (window frames, gutters, piping, wall cladding), in chemical plant and for blow moulded bottles. PPVC is used for such diverse purposes as wire covering, vinyl leathercloth and upholstery, wallpaper finishes, playballs and toys, mine belting, garden hose, and clothing. Vinyl chloride–vinyl acetate copolymers are used for flooring compositions and for gramophone records.

UNSATURATED POLYESTER RESINS

Alternative names
Polyester laminating resins

Polymer structure
There are a very large number of different polyester materials. These include fibres (such as Terylene and Dacron), thermoplastic injection moulding and blow moulding materials, thermosetting laminating and moulding compounds including dough moulding compounds (DMC) (q.v.) and sheet moulding compounds (SMC), surface coatings and rubbers. All contain the ester group $-COO-$ repeatedly in the polymer chain or network, but differ considerably in the chemical groups between the ester linkages. This section is concerned only with unsaturated polyester resins designed primarily for laminating, particularly with glass fibre. (The term *unsaturated* implies that there are double bonds in the polyester resin.)

These resins are supplied as linear polymers of low molecular weight, usually blended with styrene or some other reactive monomer to give a syrupy liquid. Mixing the resin with a peroxide (and an accelerator such as cobalt naphthenate if reaction is to proceed at room temperature) causes a reaction linking the polyester resins via styrene molecules which leads to a hard rigid network. The process is schematically indicated below:

Where

I = peroxide
     indicator
     system

S = styrene

Polymer status
Unsaturated polyester resins are *the* general purpose resin used in the manufacture of glass-reinforced plastics, widely used for example in boat building, lorry and sports car bodies, sports goods, building materials etc.

Principal variants
Whilst general purpose resins form by far the bulk of the market specialized grades include flexible resins, resins of improved heat resistance, flame retardant resins and low shrinkage resins.

Key application properties
Possibility of very simple manufacturing processes using low cost equipment. Possibility of hardening without application of external pressure or use of elevated temperatures. Comparatively low cost of resin (cf. epoxide resins). Finished products have good hardness and rigidity coupled with a high strength/weight ratio.

Selected typical properties
The data given below is for laminates made from glass fibre products with the polyester resin.

	Hand lay-up	Press formed mat laminate	Fine square woven cloth laminate	Rod from rovings
Specific gravity	1.4–1.5	1.5–1.8	c. 2.0	2.19
Tensile strength (MPa)	55–120	125–175	210–310	1030
Flexural modulus (MPa)	3440	4150	6890–13800	45000

Glass transition temperature (°C)	Cross-linked – not applicable
Crystalline melting point (°C)	Amorphous polymer – not applicable
Power factor (10^6 Hz)	<------- 0.02–0.08 ------>

Chemical and solvent resistance There are no solvents. Laminates may be damaged by strong alkalis and acids or swollen by polar solvents. Distilled water can have an adverse effect where there are exposed fibre–resin interfaces, presumably by capillary action along interface. Salt water is less damaging. Where chemical resistance is all important epoxide or furane resins may be more suitable. These two materials are not discussed in this Handbook but have been considered by the author elsewhere (Brydson, J. A. (1989) *Plastics Materials*, 5th edn. London: Butterworths.

Maximum service temperature This is very dependent on the grade of material used, the time of exposure and the stress levels involved. Short term exposure up to 200 °C is possible.

Burning properties Most grades burn with a smoky black flame. Fire retardant grades may be used but flame retarding additives are generally preferred when required.

Key processing properties
- Because of low processing temperatures any traces of hygroscopy are not normally of any importance.
- Resins are generally of low viscosity and pourable at room temperature.
- In the absence of peroxides and light, resins have good storage stability although gelation may occur on prolonged storage.
- Cross-linking is accompanied by high levels of shrinkage (up to 8%). In glass laminates this may lead to the glass fibres being raised above the level of the resin. Dilution with unreactive polymers leads to the so-called low profile resins used particularly with sheet moulding compounds.

Processing
Large quantities of polyester laminating resins are used in conjunction with glass fibre mat in hand lay-up processes. In these processes, the resin is mixed with the hardening system and then applied by brush and roller to a glass mat which has been placed in position on the mould surfaces. Hardening can occur without application of external pressure at normal ambient temperatures. A number of materials may inhibit or retard cure including water, sulphur and, with some grades, air. Modifications of the hand lay-up process include vacuum bag and air pressure assisted techniques, while matched metal moulding techniques using either fibre preforms or sheet moulding compounds give products a superior finish. Specialized techniques include filament winding (of particular use for pressure vessels and pipes) and pultrusion (for rods, including fishing rods). Higher strength laminates may be obtained using glass cloth or by the use of carbon fibres.

Applications
The resins are widely used in transport (cab bodies for both lorries and locomotives, sports car bodies, boats) and in building applications.

CHAPTER 3

SIMPLE IDENTIFICATION SYSTEM

INTRODUCTION

With several hundred types of polymer on the market, most of which themselves vary in such features as detailed structure, molecular weight and the presence or otherwise of additives such as glass fibre fillers, fire retardants, etc. a fully confirmed identification will usually require at least an infra-red analysis. There are however a number of occasions where simple tests for identification will provide adequate information.

One such example is where a company uses about twenty different materials for injection moulding and there has been a stock mix up. It is often possible to devise a simple scheme to discriminate between such materials and such a scheme is described below to embrace common injection moulding and extrusion materials. The equipment required is minimal and consists of the following:

1 source of flame, e.g. bunsen burner or blow-lamp;
2 metal spatula;
3 copper wire, one end embedded in a piece of cork so that it may be held in the hand when being heated;
4 glass beaker;
5 measuring cylinder or some device for measuring liquid volumes.

The scheme is designed to cover the following thermoplastic materials:

Polyethylene	Polypropylene
ABS	TPS
PPO	Polycarbonates
PTFE	PMMA

Polyvinyl chloride Polystyrene
Nylons Polyacetals
Poly–4–MP–1 Polysulphones

Small samples of reference materials should be available in order to compare the unknown materials, particularly with respect to burning behaviour. It should however be stressed that:

1 Burning behaviour may be modified by the presence of additives such as flame retardants.
2 The results obtained from these tests can only be indicative; particularly when the presence of polymers outside of this scheme cannot be ruled out.

PROCEDURE

1 Drop a small piece of the material into a beaker of water. If the material floats go to 2; if it sinks go to 3. (It is essential that the piece is solid and that there are no bubbles on the surface.)

2 **The sample floats on water**. While many rubbers float the only solid plastics that do so are the polyolefins such as polyethylene, polypropylene and poly-4-methyl pentene-1 (covered in this scheme) and EVA (which is not).

Drop a small piece of the material into a mixture of 62 volumes of methanol and 38 volumes of water. If the material floats it is either **polypropylene** or **poly-4-methyl pentene-1**. If the moulding is transparent then it is probably the latter, if translucent the former. If it sinks in the water–alcohol mixture (but floats on water) it is likely to be **polyethylene**. (It could also be polypropylene or poly-4-methyl pentene-1 containing a filler.) (By varying the ratio of methanol to water it will be possible to distinguish between high and low density polyethylenes. For guidance solutions of specific gravities 0.96, 0.92 and 0.91 contain 31, 57 and 62% methanol by volume respectively.)

3 **The sample sinks in water**. Heat sample in a flame, preferably using a spatula. Remove from the flame. If the sample has not burnt or ceases to burn within 10 secs go to 4; if it continues to burn go to 6.

4 **The sample does not burn or extinguishes itself within 10 seconds.** Of the materials in the list **polytetrafluoroethylene** (PTFE) is indicated. (Unfilled this material is an opaque white material which will retain its form stability well over 250 °C.)

5 **The material burns but extinguishes itself on removal from the flame.** If the flame is yellow with a green base and the odour is acrid **PVC** is indicated. Heat a piece of copper wire in a flame and press it against a sample of the material. Return the wire to the flame. A green colour to the flame indicates the presence of chlorine. Only PVC, of the materials considered here, contains this element.

If the flame is blue with a yellow tip, the odour resembles burning vegetation and the product melts to give a clear free flowing liquid that may be drawn into a fibre a **nylon** is indicated.

If the flame is almost white and emits an odour of sulphur a **polysulphone** is indicated.

6 **The material continues to burn after removal from the flame source**. If it burns without smoke – go to 7. If it burns with emission of smoke – go to 8.

7 **The material burns without smoke**

(a) The flame is a pale blue. The polymer drips as it burns and gives off a strong smell of formaldehyde on extinguishing the flame. A **polyacetal** is indicated. (Also note that polyacetals being highly crystalline are opaque even when unfilled.)
(b) The flame is yellow with a blue base. It emits a sweet odour of methyl methacrylate. **Polymethyl methacrylate** is indicated.

8 **The material burns with a smoky flame**

(a) If the material is difficult to ignite initially, burns with a yellow flame, emits a slight phenolic odour and leaves considerable amounts of char **polycarbonate** is indicated.
(b) If the polymer has a yellow flame with a blue base, is very smoky and burns with a characteristic styrene odour **polystyrene** or **toughened polystyrene** is indicated. Polystyrene mouldings emit a metallic ring when dropped onto a hard surface; toughened polystyrene mouldings do not. Where there is an acrid odour present **ABS** may be suspected. ABS also burns with a

smoky yellow flame with a blue base but the odour is characteristically acrid and pungent and should be compared with a known sample of ABS. If the odour is more reminiscent of phenol then a **PPO** derivative such as **Noryl** is indicated. Once again this should be checked against a known standard.

CONFIRMATORY TESTS AND MORE EXTENSIVE ANALYSIS

The above testing scheme can only be regarded as providing indications. It has the virtue that it can easily be adapted by a manufacturing centre to cover the range of materials it handles in conjunction with a set of standard samples. **It should not be used to identify unknown materials from an outside source, since materials other than those on the list given in the preamble to the scheme may be involved.**

For such purposes it will first be necessary to subject samples to an infra-red analysis which, in effect, provides a fingerprint for the polymer and to compare the infra-red spectrum produced with published spectra of known materials. Most infra-red spectrometers may be linked to a data base which can be loaded with a comprehensive collection of polymer spectra.

For many purposes such an analysis will itself not be complete as information may be required on components present such as fillers, plasticizers, antioxidants etc. Data may also be required on polymer variables such as molecular weight, levels of polymer branching and polymer microstructure.

Several books have been published on this subject and the following texts are particularly recommended:

Haslam J., Willis H. A., Squirrell D. C. M. (1972). *Identification and Analysis of Plastics*. 2nd edn. London: Butterworth (748 pages)
This is an excellent text, albeit somewhat dated, and includes about 300 infra-red spectra of polymers and major additives.

Urbański J., Czerwiński W., Janicka K., Majewska F., Zowall H. (1977). *Handbook of Analysis of Synthetic Polymers and Plastics*. (Translated from the Polish by G. Gordon Cameron.) Chichester: Ellis Horwood.
Another excellent text covering simple and advanced testing with an excellent collection of references but with few spectra.

Hummel D.O. (1968). *Atlas der Kunstoff-Analyse Band 1 Hochpolymere und Harze*. München: Carl Hanser Verlag.
A major collection of polymer spectra.

Saunders K. J. (1966). *The Identification of Plastics and Rubbers*. London: Chapman and Hall.
A small monograph of 54 pages describing elementary tests of identification.

Crompton T. R. (1971) *Chemical Analysis of Additives in Plastics*. Oxford: Pergamon.
162 pages devoted to an important aspect of plastics analysis.

PROCESSING

PROCESSING FUNDAMENTALS

THE SHAPING OPERATION

Plastics, and polymers in general, may be converted into products in a wide variety of ways. Whatever process is chosen it may be divided into two stages:

- getting the shape;
- setting the shape.

The shaping operation can be carried out with the polymer existing in one of the following states:

- As a melt (as in compression, injection and blow moulding and extrusion).
- In the rubbery state (as in vacuum forming).
- In solution (as when casting film or fibre spinning).
- As a suspension (as in latex technology or PVC paste processes).

- A a liquid monomer or low molecular weight polymer (as in casting and laminate production).
- As a rigid solid (in machining operations).

While in the last instance there is no special operation needed to set the shape, such a stage will be required with the other approaches. In the case of thermoplastic melts or materials in the rubbery state, the shape will be set by cooling whilst with thermosetting plastics setting will occur by cross-linking, a chemical process. With solutions it will be necessary to remove solvent while with suspensions a method will be required to gel the dispersion (in the case of PVC paste by heating, to enable plasticizer to be absorbed into the polymer particles). Liquid monomer and low molecular weight polymer will need to be polymerized and/or cross-linked.

In each process it is necessary to understand the effect of polymer properties on the process technique. This will be exemplified in the next section by reference to the melt processing of thermoplastics.

THE EFFECT OF POLYMER PROPERTIES ON PROCESS TECHNIQUE

Polymer melts may be shaped by such techniques as squeezing in a mould (compression moulding), injecting into a mould under pressure (transfer moulding and injection moulding), forcing through a hole (extrusion), or forcing through an annular hole and inflating the resultant tube in a mould (extrusion blow moulding).

When processing thermoplastics melts the following factors should be taken into account in order both to process efficiently and obtain quality products:

1 water absorption of raw materials;
2 physical form of the raw material;
3 thermal stability of the polymer;
4 flow properties of the molten plastics material;
5 adhesion of melt to metal;
6 thermal properties affecting heating and cooling of melt;
7 compressibility and shrinkage;
8 frozen-in orientation.

WATER ABSORPTION

If a polymer is allowed to absorb water, or water is simply allowed to condense on the surface of the granules, before the material is fed to the barrel of an injection moulding machine or extruder, that water will be heated to above 100 °C in the barrel. Under pressure in the barrel it may remain liquid but as the melt emerges into space from the barrel the water will turn to steam causing bubbles in the product. In many instances the bubbles will be flattened by the shearing of the melt and in a transparent plastics material will reflect and refract light giving the appearance of mica

marks or tiny slivers of metal. The existence of such imperfections may also adversely affect the physical properties of the product.

The problem is particularly serious with polymers showing high values of water absorption. For example, in the case of nylon 6 this can be as high as 14% at equilibrium. It is also serious with polymers that have to be processed at high temperatures where a small amount of water can turn into large amounts of steam. Examples of such materials are polycarbonates and polysulphones.

Even where the polymer may have a very low water absorption some additives may be sensitive and cause problems.

Good practice requires that bags of granules stored in a cold room should be brought into the converting shop some time before they are opened thus allowing the granules to come to temperature equilibrium with the surroundings. Failure to do this will encourage moisture in the atmosphere to be deposited on the granules.

Where the plastics material tends to absorb water (is hygroscopic) the following actions may be necessary (given in order of the severity of the problem):

1 Use the granules as soon as the bag is opened.
2 Dry the granules in a drying oven, or in the case of sensitive materials a vacuum drying oven, before feeding to the machine.
3 Use a hopper mounted material drier to prevent absorption or re-absorption of water in the hopper. Hot dried air may be used both to power the hopper loader and to dry the material in the hopper loader.

It should be noted that effects similar to those given by water (bubbles, mica marks) can also be given by volatiles present in the compound (e.g. plasticizers), presence of monomer (which may have been produced by degradation in the barrel, particularly in the case of acrylic polymers) and entrapped air.

PHYSICAL FORM OF THE RAW MATERIAL

Plastics materials may be fed to the shaping equipment in slab form, in powder form or as pellets.

The slab form is most useful with the calendering process (q.v.) and with compression moulding for ease of handling and, in the latter case, to reduce voids. In powder form, problems occur in most processes in avoiding entrapped air and special conditions will be required. For most other applications pellets provide the normal feedstock.

Ideally, for most processes, the pellets should have a very narrow distribution of sizes and of similar dimensions in each direction. Experiments have shown that for extrusion, highest output is obtained when the pellets are spherical but this clearly causes problems when material is split onto the floor.

Where the particles are irregular and of a wide particle size distribution, as with regrind, problems may occur in feeding to the hopper and bridging in the hopper. Particles may also be unevenly melted and air trapping may occur and extra working of the melt may be necessary.

Raw material is sometimes fed to the processing equipment in the form of powder. Potentially there is a saving in cost because of the ability to avoid the pelleting stage. The melt will also have been subjected to less heating by the time it reaches the shaping equipment. Problems do however arise because of packing and powder flow difficulties, moisture absorption and electrostatic charge generation (with the possibility of explosions). For these reasons special powder handling systems are required which are more expensive and whose cost will have to be offset against the savings made. Air entrapment may also require the use of longer barrels and screws which may not only themselves be more expensive but also lead to more working and hence total heating (heat history) of the plastics material.

THERMAL STABILITY OF THE POLYMER

A major attraction of thermoplastics materials is that scrap may be reworked. In practice however polymers do become damaged by processing at elevated temperatures. The extent of damage varies from material to material (sometimes even from grade to grade) and

processors should pay particular attention to manufacturers' recommendations on the amount of heating a material can withstand.

Polymer damage may take various forms of which the most important are the following:

1 Depolymerization to monomer. This occurs with relatively few important plastics, with polyacetals and polymethyl methacrylate being notable examples. In the former case the monomer is formaldehyde, a gas. If the polymer degrades to such an extent that large amounts of gas are formed within the barrel of the processing equipment, explosions may occur. In this case acids may speed up the degradation and serious accidents have occurred where PVC (which gives off hydrochloric acid – see below) and acetal have become mixed together and melted. In the case of the acrylic material, polymethyl methacrylate, the problem is not usually so catastrophic but the monomer may volatilize causing bubbles and other effects similar to those caused by water. (The monomer has the same boiling point as water, i.e. 100 °C.)
2 Chain scission. This is degradation through the chains breaking. Scission can occur just by intensive shearing during processing but may be made more severe by the presence of materials such as oxygen. One material which is subject to degradation in this way is polypropylene and which as a result may become more brittle.
3 Cross-linking. This is also often aggravated by the presence of oxygen. Polyethylene, in contrast to polypropylene, is affected in this way. At an early stage small amounts of gel can cause problems during manufacture of film whilst at a more advanced stage melt flow becomes more difficult.
4 Evolution of hydrochloric acid. This is a particular problem with PVC. The presence of hydrochloric acid may cause corrosion (special metals should be used for barrel and screw liners), health hazards and also speed up further degradation which will also affect colour, melt flow and physical properties of the finished product.
5 Miscellaneous chemical changes. Because of the wide range of possibilities the effects may be very diverse. Common amongst them are discolouring of the polymer and a deterioration in electrical insulation properties.

FLOW PROPERTIES OF MOLTEN PLASTICS

Melt processing of thermoplastics involves flow of the melt so the flow properties of polymer melts are important. Unfortunately, the flow behaviour of polymer melts is complex and in injection moulding the situation is aggravated by the fact that flow in the mould takes place as the melt is cooling. The section is intended to provide an introduction to the subject. The specimen calculations and data sections of the Handbook provide examples of the application of flow theory to plastics processing.

It is commonly understood that, at a given temperature, some materials flow more easily than others and that if a given tube was filled, in turn, with a series of liquids the pressure (P) required to make the material flow out of the tube at a given flow rate (Q) would vary from material to material. Plots of pressure against flow rate will give straight lines or curves whose slope will give a measure of the *viscosity* of the liquids. Such data is however quantitatively only suitable for a tube of the same dimensions and it is necessary to seek a way of expressing results more generally. This is best understood by reference to Figure 4.1. In Figure 4.1 we picture two very large plates, parallel to each other with the bottom plate stationary and the top plate moving parallel to the bottom plate at a velocity u under a force F with the space between the plates filled by the liquid under investigation. (In practice the situation is not realistic as liquid would flow out of the edges between the plates.) For many simple liquids, such as water it is found that if the force is doubled then the velocity of the top plate is doubled. More generally with such materials it is found that, if A is the area of the top plate and r is the separation between the plates

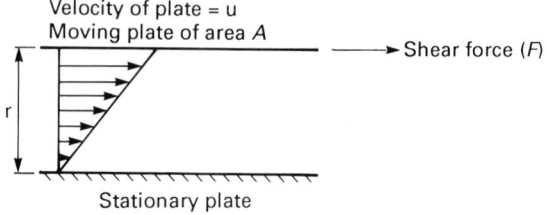

Velocity of plate = u
Moving plate of area A
→ Shear force (F)

r

Stationary plate

Figure 4.1 Shear deformation of an ideal (Newtonian) fluid. Shear stress (τ) = F/A and shear rate ($\dot{\gamma}$) = du/dr. Shear stress is directly proportional to shear rate.

$F/A = \mu\,(u/r)$ [or in calculus notation
$\mu\,(\mathrm{d}u/\mathrm{d}r)$].   (2.1)

The term F/A is known as the *shear stress* and is denoted by τ, and $\mathrm{d}u/\mathrm{d}r$ as the shear strain and denoted by $\dot{\gamma}$. The constant μ is known as the coefficient of viscosity. Equation (2.1) may hence be written

$$\tau = \mu\,\dot{\gamma}.   (2.2)$$

Such an expression is independent of the dimensions of the equipment used to measure the viscosity. Materials that obey this relationship are known as *Newtonian liquids (or fluids)*.

Polymer melts do not show such a simple straight line relationship between shear stress and shear rate, and generally are of the form shown in Figure 4.2. Such behaviour is said to be pseudoplastic. Unfortunately there is no simple equation which accurately represents polymer melts although the *power law equation*

$$\tau = K(\dot{\gamma})^{\,n}   (2.3)$$

provides an approximation. K is known as the consistency index and is a measure of viscosity while n indicates the extent which the material deviates from Newtonian behaviour and is known as the *flow behaviour index* or simply as the *flow index*. With polymer melts $n < 1$, the lower the value the more *non-Newtonian* the melt. (In the case of a Newtonian liquid $K = \mu$ and $n = 1$.)

For most calculation purposes it is better to use a *flow curve* of the type shown in Figure 4.2, the data most commonly being obtained using a *capillary rheometer*. In this case both shear stress and shear rate are zero at the centre of the tube but a maximum at the tube wall. For convenience, therefore, curves are compiled from calculations of τ and $\dot{\gamma}$ at the tube wall and given the subscript w. For a tube

$$\tau_\mathrm{w} = PR/2L,   (2.4)$$

where R and L are the radius and length of tube respectively.

The values for true shear rate are obtained from the Rabinowitsch equation which has the form

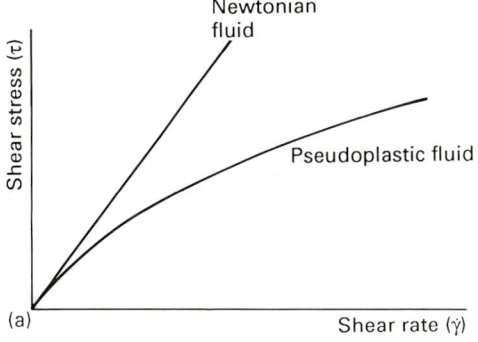

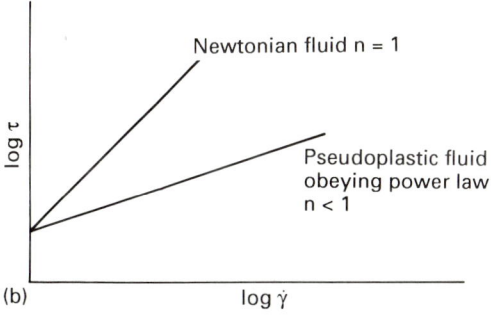

Figure 4.2 *(a) Shear stress–shear rate relationships for Newtonian and pseudoplastic fluids. (b) log τ — log γ̇ plots for Newtonian fluid and pseudoplastic fluid obeying the power law. Polymer melts are almost always pseudoplastic but only approximate in their behaviour to that of a power law fluid*

$$\dot{\gamma}_w = [3Q + P(dQ/dP)]/\pi R^3. \qquad (2.5)$$

In practice however it is just as useful in calculations to use the data for the *apparent shear rate at wall* denoted by $\dot{\gamma}_{w,a}$ and which is given by

$$\dot{\gamma}_{w,a} = 4Q/\pi R^3. \qquad (2.6)$$

The formulae for other cross-sections are given in the data and worked example sections of the handbook.

Elastic effects during flow

The flow of polymer melts differs from simple liquids in that in the former case it is necessary to move long chain molecules. These tend to move in segments rather than whole molecules at a time. Furthermore, entanglements may effectively increase the apparent molecular weight. There will also be a tendency for the molecules to orient during flow but to recoil or relax on cessation of flow. If the melt solidifies before relaxation is complete, there will be

some frozen-in orientation and the material will be anisotropic in its behaviour (i.e. it will have different properties in different directions). Effects associated with molecular orientation and (partial) recovery are generally referred to as elastic effects. Important elastic effects are:

1 die swell;
2 melt fracture;
3 sharkskin.

Die swell occurs as a result of polymer molecules uncoiling or orienting as they are sheared on passing through a die of an extruder. On emerging from the die the molecules, in the absence of continuing shear forces, tend to coil up with shrinkage in the direction of flow but expand at right angles to the flow, this resulting in the die swell. Similar effects occur on emerging from between the rolls of a calender.

The following factors increase the die swell:

1 Increasing shear (up to the critical shear rate) (see below).
2 Decreasing temperature (at constant output rate or shear rate.
3 Decreasing length of die.
4 Increasing reservoir to capillary diameter ratio to a limit of about 12:1.
5 Slit die rather than a circular die orifice.

The first three of these effects are illustrated in Figures 4.3 and 4.4.

Melt fracture is a phenomenon occurring above a critical point in the flow curve (i.e. above a critical shear stress and critical shear rate). It is characterized by some form of helical deformation of the extrudate (Figure 4.5). In the case of polypropylene and high density polyethylene a rod-like extrudate takes the form of a screw thread at low levels of melt fracture but more irregular distortions at higher levels. With polystyrene the effect is more like that of a highly extended spring. Melt fracture is most prevalent with products having a small cross-section. The origin of melt fracture, sometimes also known as elastic turbulence, remains a subject of debate, but it is a reality which has to be avoided when making commercial products. In effect this means that processing must take place below the critical point which itself depends on the following factors:

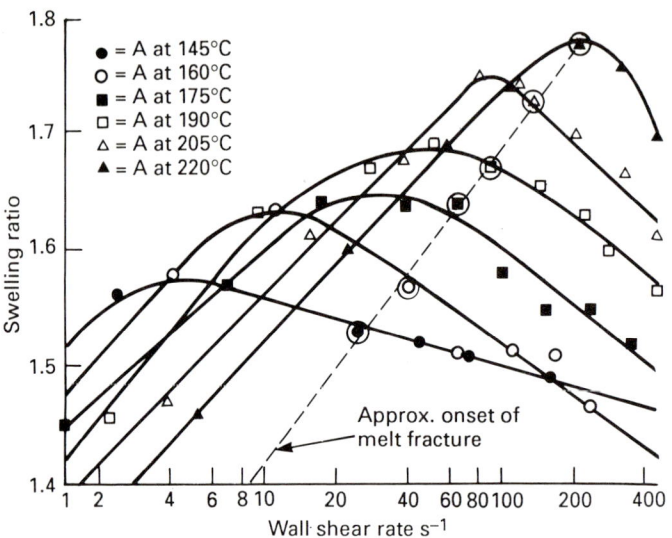

Figure 4.3 *Swelling ratio against shear rate for a low-density polyethylene at various temperatures (after Beynon & Glyde, 1960).*

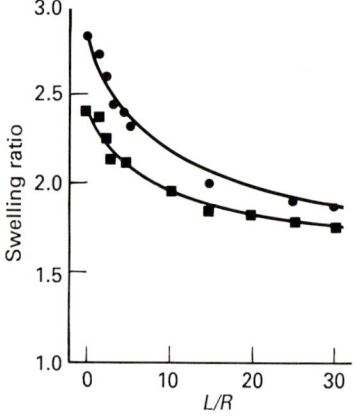

Figure 4.4 *Swelling ratio versus capillary L/R for a polyethylene of MFI 2 (after Bagley et al., 1963). Journal of Applied Polymer Science 7, 1661. ● shear rate, γ̇ = 200 s⁻¹ ■ shear rate, γ̇ = 50 s⁻¹.*

1 Increasing temperature increases both critical shear stress and critical shear rate, the latter considerably (see Figure 4.6).
2 Decreasing the molecular weight (it is commonly observed for a given polymer that the product

$$\tau_c \, \overline{M}_w = \text{constant} \qquad (2.7)$$

when $\overline{M}w$ is the weight average molecular weight and τ_c is the critical shear stress (see Figure 4.7)
3 Tapering the entry to the die parallel.
4 Slight tapering of the 'die parallel'.

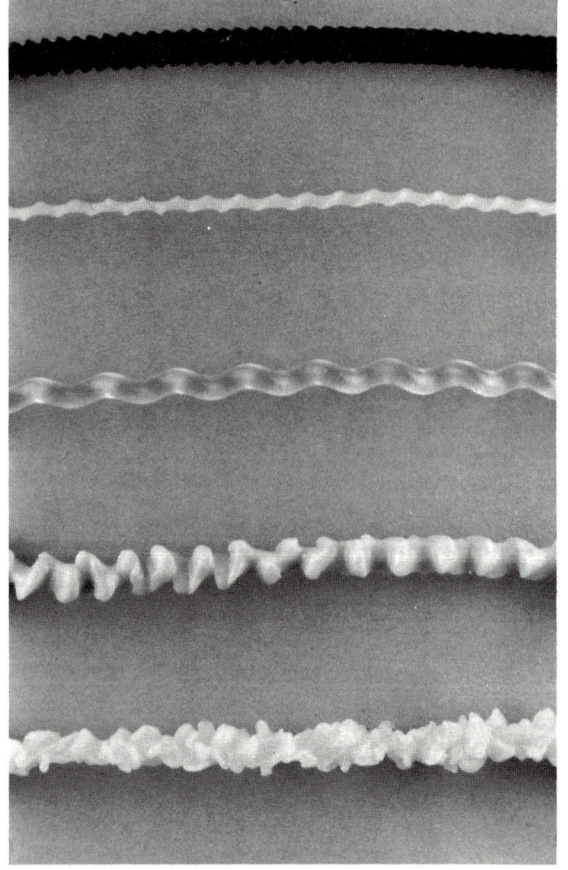

Figure 4.5 *Typical extrudates showing melt fracture. The top two examples are of an unknown polyolefin and a high density polyethylene respectively and show a regular helical distortion. The two lower samples are for the same high density polyethylene extruded at higher shear rates and show progressively less regular distortion. The central sample shows the characteristic distortion exhibited by polystyrene*

Sharkskin also gives rise to surface defects. In this case the distortions are perpendicular to the flow and not helical, taking the form of ridges. In some cases the ridges are barely visible to the naked eye while in other instances the distortion may go deep into the extrudate. It occurs above a *critical linear extrusion rate* (expressed in such units as cm/sec or ft/min) as opposed to a critical shear rate as in melt fracture. The critical linear extrusion rate, and hence the propensity to sharkskin, depends largely on the following factors:

1 The broader the molecular weight distribution of the polymer the less the tendency to sharkskin.
2 Both raising or lowering the extrusion temperature may help to alleviate sharkskin.

Sharkskin is believed to arise as the profile of the melt flow changes as it leaves the die with considerable sudden stretching of the outer layers. If the melt is weak it tears periodically giving rise to a series of perpendicular ridges. Raising the temperature helps the melt to flow and withstand the sudden stretching, while lowering the temperature at the die exit may make the melt sufficiently tough to withstand the sudden stresses that occur near the surface.

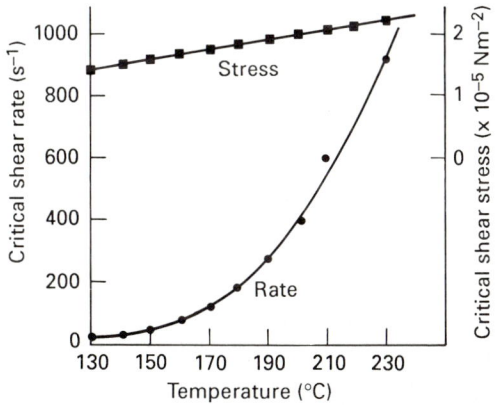

Figure 4.6 *Effect of melt temperature on onset of melt fracture (elastic turbulence) in polyethylene (after Howells & Benbow, 1962). Transactions of the Plastics Institute* **30**, 240

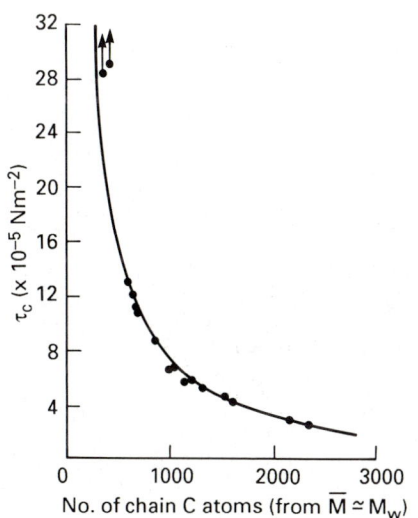

Figure 4.7 *Effect of molecular weight on critical shear stress at onset of elastic turbulence in poly(methyl methacrylate) (after Howells & Benbow, 1962). Transactions of the Plastics Institute,* **30**, 240

Flow in tension

Most studies of polymer flow have been concerned with shear flow. Recently there has been increasing awareness of the importance of tension flow and tension stresses in some processes. One example of tensile flow occurs in the stretching of a strand or filament of molten polymer. Another example of tension stresses occurring is during the injection moulding of a disc using a centre gated mould. As the melt flows into the cavity the advancing front of the melt increases in its circumference and circumferential tensile stresses are set up in the melt. This may cause some circumferential orientation to occur. This is most clearly seen with fibre-filled mouldings. Tension stresses also develop when the melt passes along converging channels.

Probably the most important effects occur in extrusion operations after the melt has emerged from the die. Such features as sag of parisons in bottle manufacture or draw down stability during film manufacture may be affected by the tension viscosity characteristics. (For further discussion see Brydson J. A. (1981). *The Flow Properties of Polymer Melts*. London: Longmans; Cogswell F. N. (1981). *Polymer Melt Rheology*. London: Longmans.)

ADHESION OF MELT TO METAL

Wetting of the polymer melt against the metal walls of processing equipment can lead to strong adhesion of polymer to metal. Amongst problems that can occur are:

1 Difficulty in removing mixes from such equipment as laboratory two-roll mixing mills. For a number of materials, such as PVC, it is common practice to use external lubricants in the formulation.
2 Where there is a strong polymer-to-metal bonding it is possible, when shutting down a machine and allowing the polymer to solidify in the equipment (such as the barrel of an injection moulding machine or extruder), for the polymer to pull pieces of metal (particularly if they are somewhat oxidized) from the barrel wall. This has been a particular problem with polycarbonates.

There is some evidence that adhesion varies with temperature and is often at a peak near the melting point of the polymer (Figure 4.8). In a single screw extruder, good pumping in the feed zone requires a higher adhesion to the barrel wall than to the screw and for this reason it is common to keep the screw much cooler than the barrel wall. In the case of PVC the adhesion is not much affected by temperature and a single screw extruder does not provide such a positive feed as a twin-screw machine.

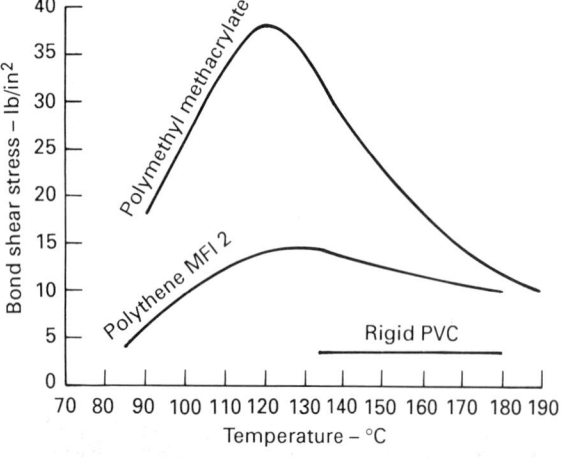

Figure 4.8 *Polymer to metal adhesion. Variation with temperature of adhesion forces of different polymers on chrome plate (from paper by L. Griffiths, in Plastics Progress – 1959. (Ed. P. Morgan). Iliffe, London (1960))*

THERMAL PROPERTIES AFFECTING HEATING AND COOLING

Different materials require different amounts of heat to raise their temperatures a specified number of degrees. It thus takes much more heat to raise a kilogramme of water from 0 to 100 °C than a kilogramme of iron. The simplest measure of such differences is the *specific heat* which is defined as the amount of heat required to raise unit mass of a material one degree. In SI units it is expressed in J/kg K.

In the case of polymer melts the situation is complicated by the fact that the specific heat varies with temperature. A further complication with polymers is that processing requires a liquid–solid phase change. Where this change involves crystalline polymers such as the nylons and polyethylenes there is a *latent heat of melting or fusion* which occurs over a narrow range of temperature and in SI units may be expressed in J/kg. Where available it is often better to use *enthalpy* curves which measure the total heat per unit weight required in heating the polymer from room temperature to the melt temperature, or to be removed when cooling the melt from its melt temperature to the mould temperature. Estimates of the enthalpy (total heat) to raise 1 kg of a polymer from an ambient temperature of 20 °C to a melt temperature required for injection moulding are given in the following table (Table 4.1).

Table 4.1

Polymer	Process temperature (°C)	Enthalpy (kJ/kg)
Polystyrene	200	310
Low-density PE	200	500
High-density PE	260	810
Polypropylene	260	670

Note: More comprehensive data for a wider range of materials is given on page 179.

This data clearly shows considerable differences between polymers. Not only will more heat have to be put into bringing high density polyethylene up to its processing temperature than is necessary for polystyrene but much more heat will have to be removed before the moulding cools down to the ambient temperature (see also Figure 4.9).

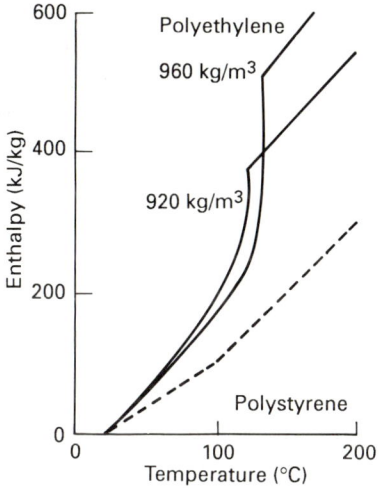

Figure 4.9 *Enthalpy versus temperature (data courtesy Hoechst)*

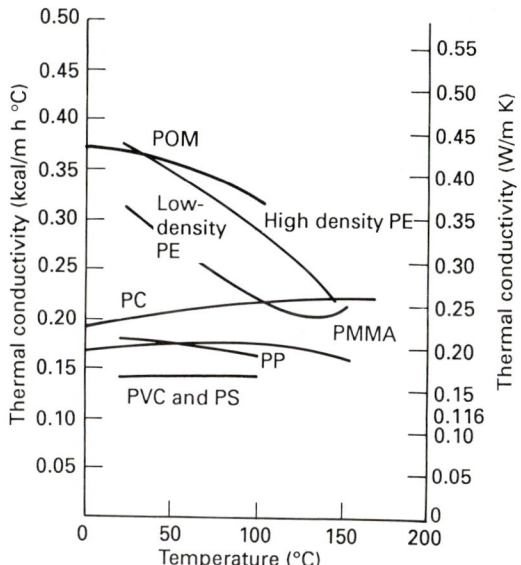

Figure 4.10 *Thermal conductivity versus temperature (data courtesy Hoechst)*

When cooling is dominated by conduction heat transfer, the rate of cooling is also affected by other material properties, namely thermal conductivity (k) and density (ρ). Theoretical analysis shows that the rate of cooling is controlled by the thermal diffusivity (α), a composite property given by the expression

$$\alpha = k/s\rho, \qquad (2.8)$$

where s is the specific heat.

The situation is complicated by the fact that although the thermal conductivity of amorphous plastics is almost independent of temperature this is not true with crystalline thermoplastics. With such materials the thermal conductivity can drop considerably between normal room temperature and the melting point (nearly 50% in the case of polyethylenes) (Figure 4.10). Examples of the use of thermal diffusivity data for assessing cooling rates in moulds are given in the section on specimen calculations (Appendix 1).

COOLING SHRINKAGE AND COMPRESSIBILITY

When polymers are in the molten state the vibrations of the molecules result in the polymer chains being pushed apart so that the volume occupied by a given polymer mass is higher than when the material is solid. The difference between melt and solid density can be quite marked, particularly with crystalline

thermoplastics, and this will be a major influence on the moulding shrinkage. Some typical data is given in Table 4.2.

Table 4.2

Polymer	Density at 20 °C (g/cc)	Density at process temperature (g/cc)
LDPE	0.923	0.746 (210 °C)
Polypropylene	0.905	0.765 (210 °C)
PMMA	1.180	1.105 (210 °C)
Polysulphone	1.370	1.230 (350 °C)
Plasticized PVC	1.480	1.390 (190 °C)

Although moulding shrinkage is usually much higher with crystalline thermoplastics than amorphous thermoplastics, moulding shrinkage is much less than the 20% indicated for cooling shrinkage in the case of LDPE. This is because polymer melts are *compressible*. At 20 000 p.s.i. the compressibility for amorphous plastics such as polystyrene is of the order of 10% and may be higher with crystalline polymers.

FROZEN-IN ORIENTATION

When polymer melts are being shaped by processes such as injection moulding and extrusion, the long polymer chains tend to be elongated or uncoiled in the general direction of flow. After shaping, the melt is usually cooled quite rapidly and there is seldom time for the oriented molecules to return to a randomly coiled-up shape by the process known as relaxation. Some orientation is thus *frozen-in* the product. Many properties vary with the direction of orientation and the product is said to be *anisotropic*. Such anisotropy may be particularly marked with injection mouldings particularly where high injection rates and low melt and mould temperatures are being used. With crystalline polymers both crystals and polymer chain may be oriented. Synthetic fibres may be considered as extreme cases of orientation and are products in which along-axis tensile strengths may be ten times that of unoriented polymer.

Care must be used when interpreting impact data. On page 11 of Chapter 1 it was demonstrated that in a rod-shaped end-gated moulding a high level of orientation led to a high value of impact strength but that with a tile-shaped mould high orientation led to low impact strength.

5

MAJOR PROCESSING METHODS

INJECTION MOULDING

In many ways injection moulding may be regarded as the embodiment of the archetypal plastics processing method. It has considerable potential because of the following features:

- The ability to operate the process as a highly automated mass production operation.
- The ability, where necessary, to make highly complex shapes in a single operation.
- The ability to make articles repetitively with consistent reproducible properties.
- The ability to make products with widely differing properties according to the choice of materials and the processing conditions.

In order to operate in such a way that the above objectives are achieved will however require particular attention to the following:

1 The selection of a suitable machine of capacity and level of operational control appropriate to the job in hand.
2 Correct selection of plastics materials.
3 Good mould design.
4 Correct machine setting and proper handling of the machine, moulds, materials and mouldings.

In principle, injection moulding consists of feeding plastics material into a chamber in which the material is heated until it is molten, and then injecting the molten material into a mould in which the melt hardens. With thermoplastics materials this is achieved by cooling in the mould to below the glass transition temperature in the case of amorphous thermoplastics and below the crystalline melting point in the case of crystalline thermoplastics. With the thermosetting plastics this hardening is achieved by chemical reactions, usually in a hot mould, which lead to cross-linking (see Chapter 1). A schematic diagram of a basic machine is shown in Figure 5.1(a). For simplicity, details of the methods used for clamping the two mould halves together are not shown.

Early injection moulding machines based on the system shown in Figure 5.1(a), suffered from the difficulty of reconciling the temperature and pressure control requirements necessary for good moulding practice. In the design shown it can be appreciated that materials in the centre of the cylinder will be at some distance from the heaters and because of the very low thermal conductivity, and hence thermal diffusivity of polymers (see Chapter 4), an excessive time will be required to heat the material in the core. There will also be a wide variation in temperature throughout the melt and this could lead to variability in the properties of the moulding. In the case of heat sensitive materials such as unplasticized PVC the layers of polymer close to the cylinder wall

often tended to decompose before the polymer at the axis of the cylinder had reached processing temperature.

At an early stage in the development of the injection moulding machine it became common practice to insert a solid cylinder of metal into the front of the injection cylinder (known as a torpedo). With such a system, shown in Figure 5.1(b), the distances that heat had to travel through the plastics material was very much reduced. Heating was much faster and temperature and product variability thus reduced.

It is however much more difficult to push material through long narrow channels rather than short wide ones and considerable pressure losses were observed between the front of the injection ram and the distant corners of the mould cavity. This was both wasteful of energy and also a factor that reduced control over the process.

In the years immediately following the Second World War many attempts were made to develop systems of cylinders and torpedoes which offered acceptable compromises between the temperature and pressure require-

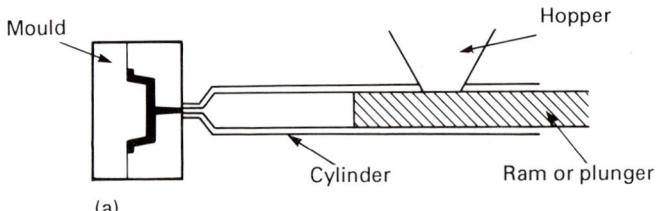

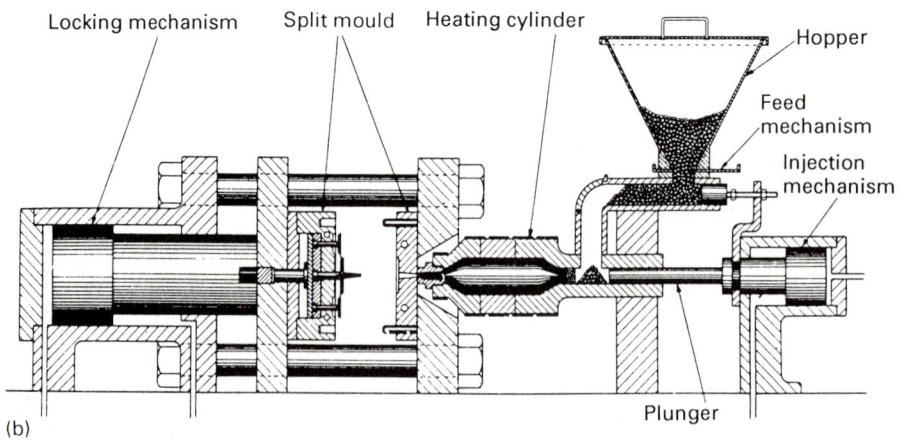

Figure 5.1 (a) Schematic outline of basic ram injection moulding machine. (b) Ram machine incorporating a torpedo or spreader inside the heating cylinder

ments. It became recognized that a key element in the solution of the problem was to separate the plasticizing and the injection stages and so-called preplasticizing machines were developed. Of the many types developed, the *in-line single screw injection moulding machine* has been so successful that it now dominates the market for injection moulding of plastics and further discussion in this section will be restricted to such machines.

The principle of the machine is seen in Figure 5.2. Granules are fed through a hopper to the cylinder and delivered up the cylinder by means of a rotating screw. In the movement up the barrel the polymer is heated, both by conduction from external heaters and by internal frictional heating as it is pumped up by the screw. There is also some mixing and homogenization of the melt caused by the rotation of the screw. The softened or 'plasticized'

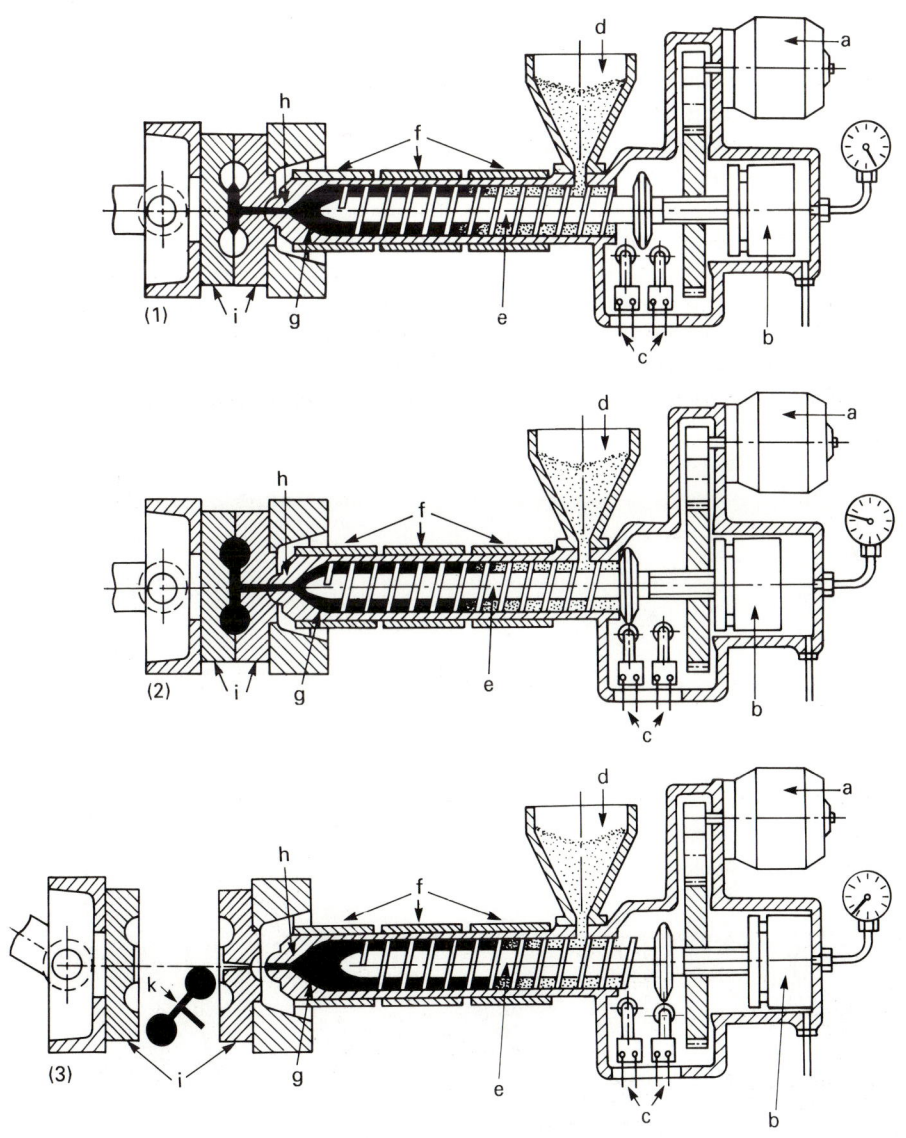

Figure 5.2 Screw injection moulding cycle. (a) Motor for rotation of the screw. (b) Drive for reciprocating movement of the screw. (c) Left limit switch for setting holding pressure, right limit switch for limiting the screw stroke. (d) Feed hopper. (e) Conveying, plastifying and injecting screw. (f) Heater bands. (g) Hot plastified material. (h) Injection nozzle. (i) Cooled or heated mould with sprue bush (attached on the locking unit). (k) Moulded part with sprue. Top: Molten polymer being injected into mould by screw acting as a ram (usually non-rotating) under injection pressure. Middle: Screw in forward position under holding pressure until the material in the gate (point of entry into the mould cavity) has frozen. Bottom: Screw has rotated and moved axially away from mould to charge sufficient material to front of cylinder for next shot. Mould opens when moulding is sufficiently hard

material is collected at the front of the cylinder and to accommodate the accumulating molten mass arrangements are made for the screw to retract along its axis until there is the correct volume of melt at the head of the cylinder. At this stage movement of the screw, both rotationally and axially, ceases. By this time the mould will have been closed and the nozzle of the injection unit will have come into contact with the mould. The screw is then activated to operate as a plunger, moving forward pushing the melt into the mould cavity. The screw is held in the forward position under pressure until the material in the gate freezes, thus isolating the material in the mould cavity from the operations of the injection unit. At this point the screw again commences to rotate, feeding more melt to the front of the cylinder. This material is held until the moulding has cooled and been ejected. Injection of the melt then takes place and the process is repeated.

In this process, plasticization and injection stages are separated so that both good heat and pressure transmission may be achieved. Furthermore the in-line screw process has the important added advantage that the system is self-cleaning, the first material into the system being, to an approximation, first out, so that problems of hold up and stagnation of heat sensitive materials are minimized.

One possible disadvantage of the process is that as the screw moves forward to inject polymer into the mould, material may leak back up the screw channel (and even over the screw flights). With most polymers this is not a problem but with some melts it may be desirable to incorporate a sealing ring into the system.

THE INJECTION CYCLE

A typical injection cycle is illustrated in Figure 5.3. The first stage, *mould closing* should be as fast as possible consistent with the final part of the closing operation not being so fast as to cause damage to the mould. There should also be time for any safety switches to become effective if necessary.

As soon as the mould is closed the screw can start to move forward and the *mould filling* stage

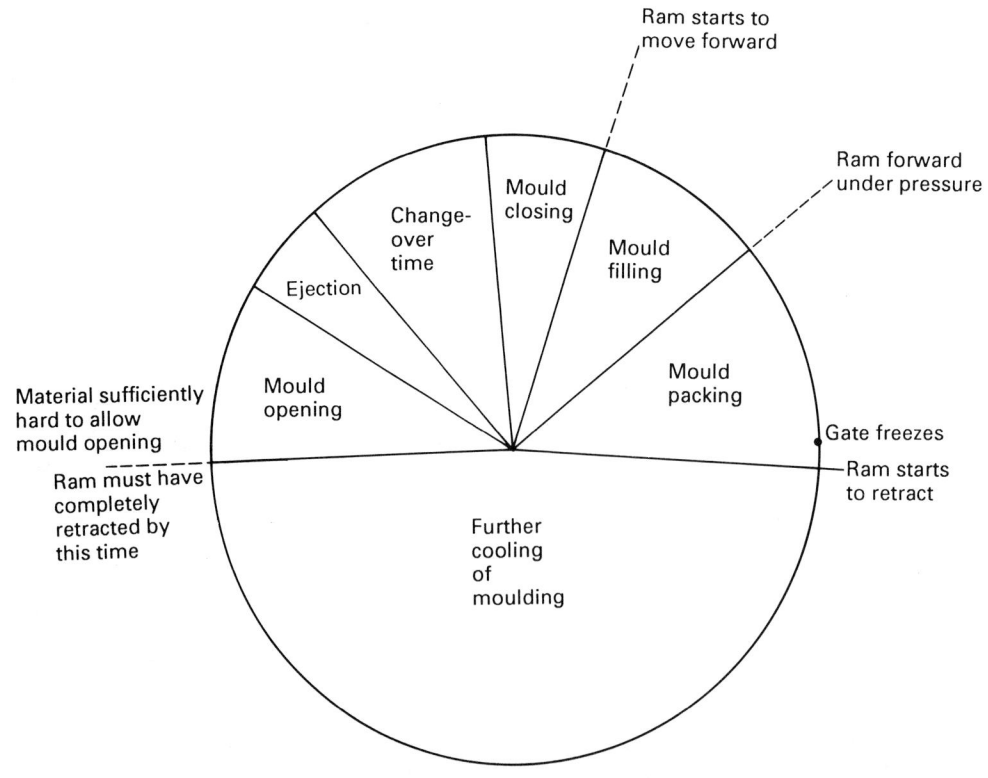

Figure 5.3 *The injection moulding cycle*

commences. When the ram is fully forward and the mould cavity is filled the pressure on the melt will increase and this will help to consolidate the material in the cavity. This is known as the *mould packing* stage. During this stage the gate freezes. When this happens any activity in the injection cylinder has no bearing on what is happening in the mould cavity.

Once the gate has frozen, the screw may be retracted whilst further cooling of the moulding occurs. Whilst this is often known as the *mould cooling stage* the melt is actually beginning to cool from the moment that it leaves the injection cylinder.

When the moulding is sufficiently solidified *mould opening* can take place, and the moulding *ejected*. There may also be a requirement for some *change-over time* before commencing the next cycle, for such purposes as cleaning or checking mould surfaces or placing inserts. It is also important to appreciate that the *screw retraction* operation must be complete before the start of the next cycle.

It is instructive to consider the pressure profile that develops in the mould cavity during the injection cycle and this is illustrated in Figure 5.4. In Figure 5.4(a) the pressure registered at the blind end of the mould cavity (the end furthest from the gate) is zero until the moment that the cavity fills. (There will of course be a pressure exerted by the front of the screw against the polymer melt in order to push the melt through the injection unit and the sprues, runners, gates and cavity of the injection mould, but until the moment of mould filling this will not be recorded by any measurements taken of pressures in the mould cavity.) Pressure build-up once filling has occurred is rapid and a high pressure value is maintained until the material in the gate freezes. As the mass of polymer in the cavity cools, the molecules become less active and exert less pressure against each other. In theory, if the mould setting has been correct, this cavity pressure should drop to zero at the moment that the melt in the cavity solidifies. In practice not all of the melt solidifies at the same moment.

If, on the other hand, the injection pressure has been too high there may be some residual pressure after the moulding has cooled. Since pressure acts in all directions, the residual sideways pressure may make ejection difficult as

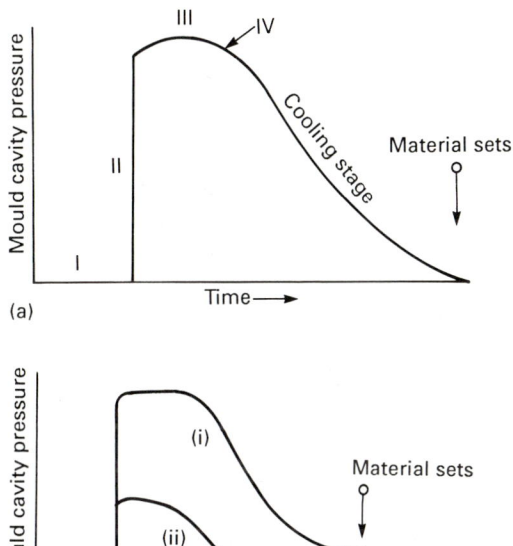

Figure 5.4 (a) Preferred trace of mould cavity pressure with time during injection moulding cycle. Cavity pressure should drop to zero at the moment when moulding has reached setting point. I – mould filling, II – pressure build-up, III – consolidation, IV – gate freezing point. (b) Undesirable cavity pressure–time traces. (i) Residual pressure when the moulding sets causes sticking problems. (ii) Cavity pressure is at zero while the polymer is still molten. Further cooling leads to sink marks, voids and shrinkage defects. Copied from Brydson, Flow Properties of Polymer Melts, p. 129, Figures 7.6 and 7.7

the moulding binds or sticks in the cavity (Figure 5.4(b)(i)).

At the other extreme, if too low an injection pressure has been employed the cavity pressure may drop to zero whilst the polymer is still molten. Further cooling may lead to contraction and shrinkage of the moulding which will exhibit features such as sink marks and voids (Figure 5.4(b)(ii)).

VARIABLE SPEED AND PRESSURE CONTROLS DURING MOULDING CYCLE

With modern injection moulding machines it is usually possible to vary the speed of the screw acting as a ram (the injection rate) during the mould filling (injection) stage. Figure 5.5 shows a moulding where such control may be beneficial and Figure 5.6 a possible injection rate – screw position profile.

In the first substage the ram will be at a high speed (V_1). This will help to push the back-flow

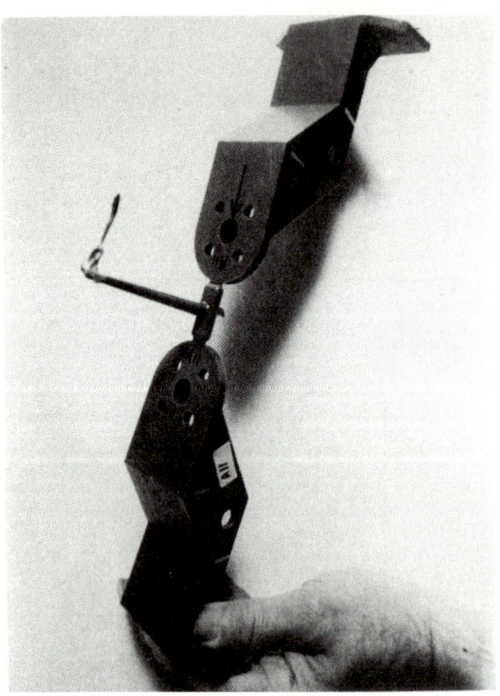

Figure 5.5 *Moulding prepared using variable speed and pressure controls. Black arrows point to holes. Reproduced with permission from Morton-Jones and Ellis, Polymer Products – Design, Materials and Processing, p. 15, plate 1*

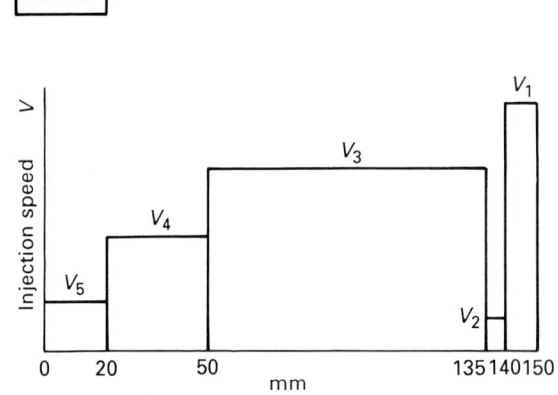

Figure 5.6 *Variation of injection rate with screw position. Copied from D. H. Morton-Jones and J. W. Ellis, Polymer Products, p. 27, Figure 2.2*

check valve onto its seating and prevent back flow down the screw channel as the screw moves forward as a ram. It also allows rapid filling of runner and sprue.

When the melt front reaches the gate the speed is reduced (V_2) to reduce the chances of jetting of the melt into the cavity. (Such jetting may cause blemishes particularly with trans-

parent mouldings.) Once the front is into the mould cavity the injection rate may be increased (V_3) to fill the bulk of the cavity as rapidly as possible.

Towards the blind end of the cavity there are some mould cores. If melt moves through the narrow gap between core and cavity wall too fast, jetting may occur; on the other hand if the movement round the core is too slow the melt fronts may not reunite sufficiently on the far side of the core and a weak *weld line* may result. On balance a reduced speed (V_4) is used.

As the melt front approaches the blind end of the cavity deceleration is programmed (V_5) to give time for air to escape from the cavity. If air is trapped and suddenly compressed it heats up to such an extent that it may degrade the polymer and cause *burn marks*.

In the filling stage the speed (rate) varies with distance (i.e. screw position). Although during this stage the pressure exerted by the screw on the melt will be varying, it will not be the pressure that is the controlling variable. Once however the mould is filled it becomes important to control *pressure* and vary this with time. Such a packing pressure–time profile is shown in Figure 5.7.

In the first stage of this cycle ($T1$) the gate is still open. A low pressure is used to reduce the probability of flashing before the skin is frozen. The main packing occurs during stage ($T2$) which is carried out at higher pressures to ensure well-formed mouldings free of sink marks and voids. In the third stage ($T3$) there is some reduction in pressure to reduce overpacking and hence excess stressing around the gate prior to the gate freezing. In the final stage, some pressure is maintained to hold the ram forward until commencement of the plasticizing stage.

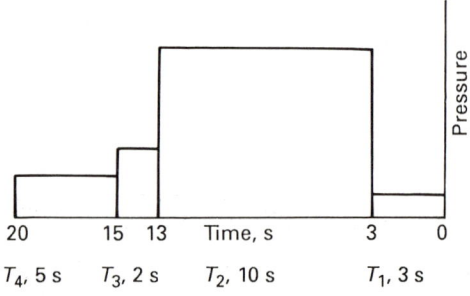

Figure 5.7 *Variation of pressure during packing stage of injection cycle. Copied from D. H. Morton-Jones and J. W. Ellis, Polymer Products, p. 29, Figure 2.4*

THE SCREW CUSHION

When setting the injection unit it is important that there is a little more melt at the front of the barrel than is required for the moulding shot. This will provide a cushion to prevent the tip of the screw bottoming on to the injection nozzle. If no such cushion is provided there will be no follow-up pressure on the melt in the mould cavity as the moulding cools, while additionally there will be excessive wear on nozzle and screw tip (see Figure 5.8).

There is no need for the cushion to be larger than necessary to ensure its existence and a 3 mm gap is commonly recommended between screw and nozzle when the screw is fully forward. If the cushion is excessive it will reduce the effective length of the screw (the distance between screw tip and hopper).

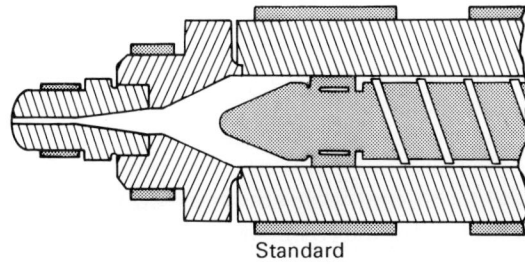
Standard

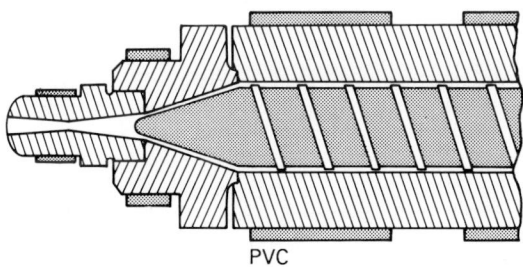
PVC

Figure 5.8 Screw head: nozzle relationships for a standard screw with check valve (see page 80) and for a screw for use with PVC. There should always be a 'screw cushion' between screw head and nozzle recess when the screw is fully forward (usually about 3 mm)

THE INJECTION UNIT

Before a discussion of the constructional features of the unit it will be useful to list the principal specifications.

Injection capacity: This is the maximum amount of material that may be injected per cycle of operations (i.e. per shot). Previously, machines were rated in terms of the weight of polystyrene that could be injected per shot. The disadvantage of this approach is that it was necessary to recalculate the weight for polymers of different densities and more recent practice is to quote the swept volume (usually in ccs) that the injection unit can deliver in one stroke at a nominated injection pressure (100 MPa is one important standard).

Length:diameter ratio (L:D ratio): This is only relevant to screw machines. The figure quoted is not necessarily the actual L:D ratio but rather is based on the effective length of the screw, which is the distance between the rear of the hopper when the screw is in the fully forward position to the end of the screw flights. It does not normally include the shut-off valve. A typical screw will have an L:D of 20:1, with higher values for vented screw units (q.v). Since a larger diameter of screw allows more contact with the polymer per turn of the screw thread, an increase in screw diameter may allow lower L:D ratios to be used. For example, whereas a typical 70 mm diameter screw will have an L:D of 18:1, a 90 mm diameter screw may have an effective L:D of 14:1.

Screw stroke. This is the maximum linear movement of the screw that can take place during the plasticizing stage. As the screw moves back the effective L:D ratio decreases. If the screw stroke is excessive the effective L:D will become so low that it will only pump poorly mixed and melted material to the head of the barrel during plasticization. To avoid any serious reduction in melt quality the maximum screw stroke is usually < 3.5D. Because of difficulties in accurate metering, the unit should not be used for moulding operations requiring only a small fraction of the maximum stroke (e.g. less than 10%). It is also generally preferable to operate below the maximum where possible.

Plasticizing capacity: This is the maximum amount of material that may be plasticized to an acceptable melt quality per unit time. This will be very dependent on the material used and is most commonly expressed in terms of weight of polystyrene plasticized per hour.

Plasticizing capacity will also be influenced by shot size, injection rate, screw design, barrel specification and melt temperature. Comparison between machines therefore requires that standard procedures, such as those of the Society of the Plastics Industry Inc. (SPI) or the European Committee of Machinery Manufacturers for the Plastics and Rubber Industries (Euromap), be employed.

Injection Rate: This is the maximum rate at which the screw or ram can eject melt from the barrel during a single shot. It may be expressed either in terms of volume per unit time or weight per unit time. It will depend on the material used. The rate is usually measured where there is no mould (i.e. when an *air shot* is made), conditions that give the highest value. In this test both the injection rate valve and the line pressure are set to a maximum.

Injection Pressure: This is the maximum pressure that may be exerted by the screw on the molten polymer during the mould packing stage of the moulding cycle. It is not

1 The maximum line pressure of the hydraulic oil between the pump and the back of the screw, nor is it
2 The maximum pressure exerted by the screw on the melt before the mould cavity is filled.

Although some machines may operate up to 5000 MPa, in practice operation should be carried out at the lowest possible pressure consistent with the required injection rate.

FEATURES OF AN INJECTION UNIT

As in the previous section the comments below apply to in-line single screw injection moulding machines.

Barrel (cylinder; plasticizing chamber): This is a robust steel cylinder which may be nitrided. Special steels or platings may be required when processing polymers that emit corrosive materials when heated, such as PVC. Since pressures used in moulding can be as high as 240 MPa ($\simeq$ 35 000 p.s.i.), pressures of the same order as developed when firing a naval gun, it is important that the walls be thick enough to withstand this pressure.

Screws: A variety of screw designs are available, selection depending on the type of material to be processed. The principles involved are largely the same as those associated in extrusion, where these variations will be discussed further. However where a screw is to be used for injection moulding a check valve such as that indicated in Figure 5.9 may be fitted to the end of the screw. This check valve (also known as the back flow valve or a non-return valve) is used to prevent melt flow back up the channel as the screw moves forward during the injection stroke. Whilst such back flow will be resisted by materials already in the channel it may be significant with low viscosity melt materials (e.g. some polyolefins and nylons). The design is such that as the screw is pumping material to the front of the cylinder during the

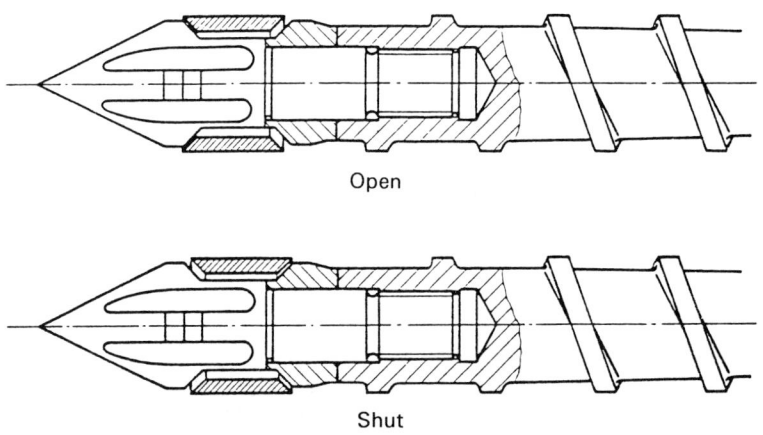

Open

Shut

Figure 5.9 *Ring check valve. Reproduced by courtesy of PPITB*

plasticizing stage the ring is pushed forward allowing material to flow past it. When the screw is acting as an injection ram the valve ring moves backwards, sealing off the screw channels from the reservoir of material at the head of the barrel.

Nozzle: This provides the link between the injection unit and the mould. To ensure a good seal the nozzle tip radius is usually a little less than the sprue bush radius as indicated in Figure 5.10. This figure shows a typical nozzle for a typical polymer melt. If, however, a low viscosity melt is being processed there is a problem of leakage or 'drooling' out of the injection unit through the nozzle at inappropriate parts of the injection cycle. This has led to a range of restricted flow valves. These may operate simply by careful temperature control of a nozzle with constricted flow, or some sort of device in which a spring

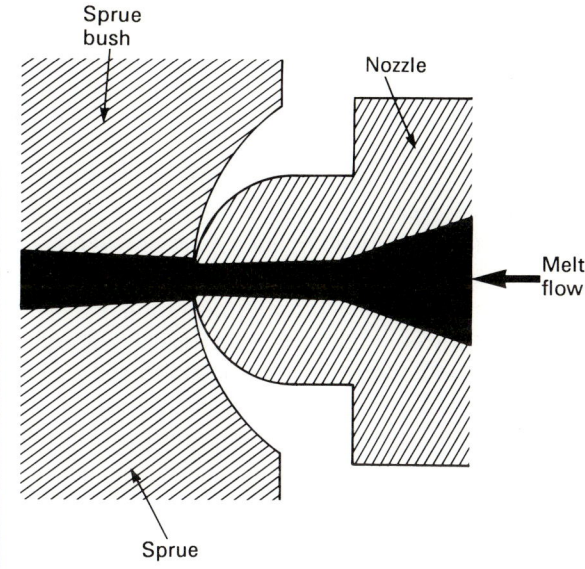

Figure 5.10 *Nozzle–sprue bush relationship. The design shown will provide good sealing and prevent material hold-up. Reproduced by courtesy of PPITB. Copied from PPITB Standards Manual, Injection Moulding, p. M4–3.1*

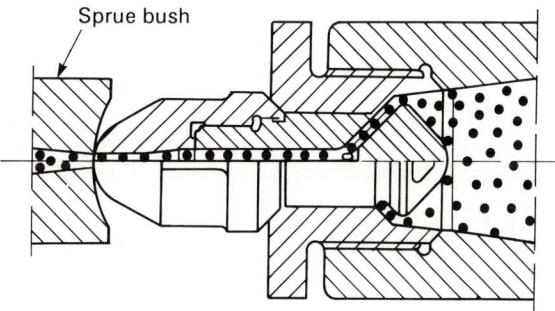

The springless type shut off nozzle relies on the contact between the nozzle and the sprue bush to open it, as the cylinder carriage moves forward.

Carriage forward –
nozzle valve open

At the end of the shot, the carriage moves back, to break the sprue with a gap between nozzle and sprue bush, the melt pressure in the cylinder closes the nozzle valve.

The springless (or self-sealing nozzle is more difficult to clean because of its melt channel complexity.

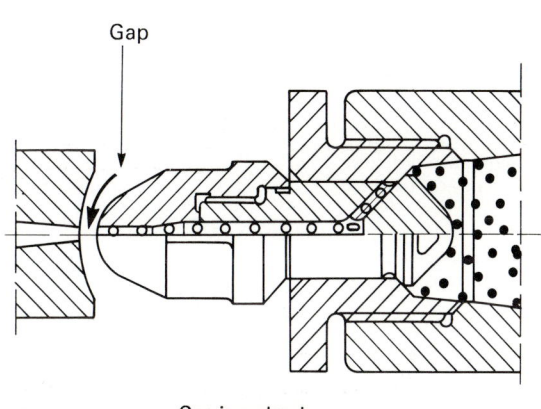

Carriage back –
nozzle valve closed

Figure 5.11 *Springless type of shut-off nozzle. Reproduced by courtesy of PPITB*

loaded sealing mechanism becomes operative at the moment that the nozzle moves away from the screw bush. Three simple designs are shown in Figures 5.11 and 5.12

Heating systems (including temperature control methods): These are a crucial aspect of the injection unit. While fluid heating (oil, hot water, steam) may be used for thermosetting plastics and vulcanizable rubbers, electrical resistance heating is standard for thermoplastics. To obtain, consistently, good quality mouldings it is important to have good temperature control systems. Early machines used simple energy regulators which either controlled, manually, the supply of current to the heaters (as with a rheostat), or controlled the proportion of time the unit was supplying heat (on-off time proportional switches). In the latter case the percentage of time that the heaters are on is controlled by turning a knob. When the regulator is 'on' full power is supplied. By means of manual trial and error sufficient heat energy is supplied to attain the desired temperature. Like a typical oven hot plate there is no actual control of temperature and precise temperature control is difficult to achieve.

The lower viscosity materials require a shut-off valve to stop the nozzle drool during the preparation of the next shot.

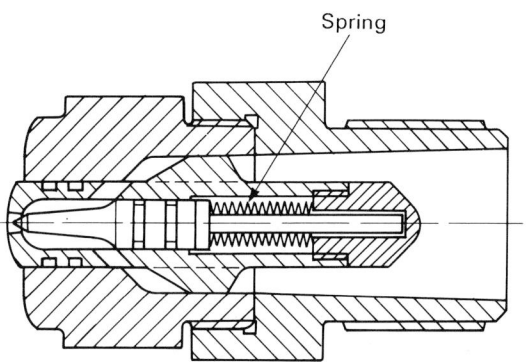

'Fuchslocher' design shut-off nozzle

Various designs of shut-off nozzles rely on coil springs to close them. As the carriage moves forward, the spring is compressed and the nozzle opened by the contact between the nozzle and the sprue bush.

These spring-loaded designs are difficult to clean and tend to become less efficient as their springs weaken under prolonged exposure to process temperatures.

Sliding shut-off nozzle with spring-loaded needle valve

Figure 5.12 *Typical spring-loaded shut-off nozzles. Reproduced by courtesy of PPITB*

These energy regulators were replaced by simple on-off indicating systems, actuated by a simple thermocouple embedded close to the inner barrel wall (Figure 5.13(a)). In such systems, the desired temperature is set and full power is supplied, so long as the measured temperature is below the set temperature. When the desired temperature is reached, the heater switches off. However, because of the finite time taken for the heat to flow from the heater past the thermocouple to the melt, a temperature overshoot will occur before the temperature drops back and the heater is switched on again. Then there is a delay whilst the heat travels through, during which time the melt will be below temperature. This process is repeated cyclically leading to the hunting effect shown in Figure 5.13(b).

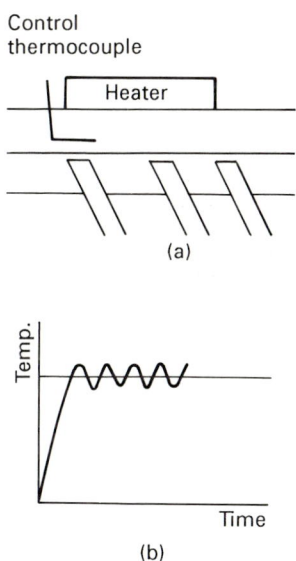

Figure 5.13 *Hunting effect (b) obtained using simple on-off controller actuated by deep-seated control thermocouple*

Better control may be obtained by using *proportional controllers*. In a simple system the control switches the power on and off as the cylinder warms. When the system is 'on' full power is supplied but as the set temperature is approached so the proportion of time that the heater is on will decrease. Such a system anticipates the approach to the set temperature but some overshooting is still possible. In the proportional-plus-derivative (PD) system the controller may additionally adjust the output signal to the heater in proportion to the rate of change of the signal. More sophisticated still are the proportional–plus–integral–plus–derivative (PID) systems which give even better control.

Where it is desired to make changes in melt temperature, external adjustment using the temperature control system may take up to several hours to settle down after a change. Rapid adjustments are more easily achieved by:

1 Adjusting screw speed. An increase in screw speed increases shear and frictional heating, raising the temperature of the melt.
2 Increasing screw back pressure (the resistance to the retraction of the screw during the plasticization stage). This also generates more frictional heat.

Cooling: Facilities are also required for the cooling of the injection unit. Hopper block cooling is necessary to prevent blocking of the feed of granules to the cylinder whilst it has also been found necessary to maintain hydraulic oil temperatures to achieve consistency in operation.

THE LOCKING UNIT

The function of the locking unit is to hold the mould closed during the mould filling and packing stages of the injection cycle. In order to perform this function large locking forces may be necessary. This is because when the cavity is filled with molten polymer, a high pressure is transmitted by the injection ram or screw onto the melt. This pressure is transmitted throughout the melt and is exerted in all directions. Of particular importance is the pressure exerted per unit area of the *projected area* of the mould cavity (the area inclusive of runners and sprues perpendicular to the direction of the locking force). This locking force must exceed the sum of the pressures on each unit area. Failure to do so may cause the mould to open and flashing will occur. In practice, the force tending to open the mould will be somewhat less than the product of injection pressure and projected mould area.

The locking unit has a number of requirements including:

1 It should be fast acting.

2 The final stage of mould closing should not be so fast that there is damage to the mould faces.
3 Highly pressurized hydraulic oil should be used as sparingly as possible.

Many types of locking unit have been devised most of which can be classed into two types, namely:

1 Direct hydraulic locking systems.
2 Toggle systems.

The basic direct hydraulic system is shown in Figure 5.14. This system uses a large diameter hydraulic cylinder mounted in the rear platen. The moving platen of the mould is attached to the cylinder ram which both opens and closes the mould and provides the clamping force. The advantages of the system are that the cylinder is self-lubricating and the system has few moving parts. In the direct system, the ram speed is controlled by varying the oil flow rate. In contrast to the toggle system variations in mould heights are easily accommodated. The main disadvantage is that it is frequently necessary to move around large quantities of pressurized oil which is wasteful both in terms of power consumption and time. It may also make the process undesirably noisy. Many systems have been developed which avoid the use of such large quantities of oil for the bulk of the closing operation; the high pressure oil being largely used for locking.

The alternative system is one involving toggle locking, a simple form of which is shown in Figure 5.15. By means of the toggle, rapid mould closing may be effected by a small locking cylinder. However as the toggle bars come into line, the rate of movement of the mould platen slows down to protect the mould surfaces from damage. If the unit has been properly set only a small force is required to keep the bars in position to provide a high clamping force. The main disadvantages of toggle systems are greater susceptibility to wear and tear of moving parts and the greater care needed to set the system for the correct mould height. Some systems exist that have characteristics of both direct hydraulic and toggle mechanisms.

The development of computer control of injection moulding has facilitated much greater control over every stage of the moulding process. As was shown on page 78, it is useful to be able to vary injection rate during the injection stroke and cavity pressure during the cooling cycle. Similarly it may be useful to vary line pressure during mould filling and vary closing speeds for the locking unit. This is quite routine for computer controlled equipment. In addition, many machines have a memory for a limited number of settings, while additional programmes may be activated by use of a programme card, tape or disc. The use of robots as an aid to removal of mouldings is now also quite common practice.

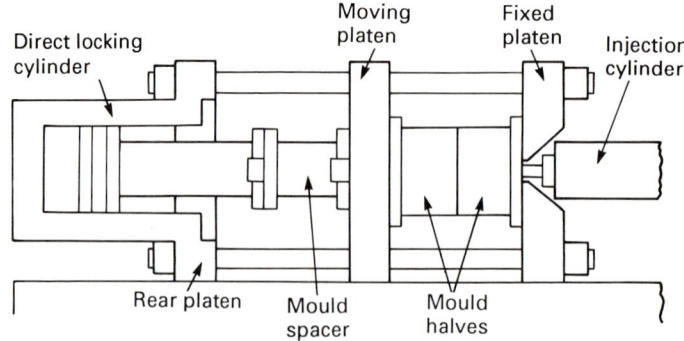

Figure 5.14 *Simple direct hydraulic locking system. The design illustrated incorporates a mould spacer which is useful in reducing the length of stroke of the ram, which is limited to the mould opening necessary to eject the moulding. Reproduced by courtesy of PPITB*

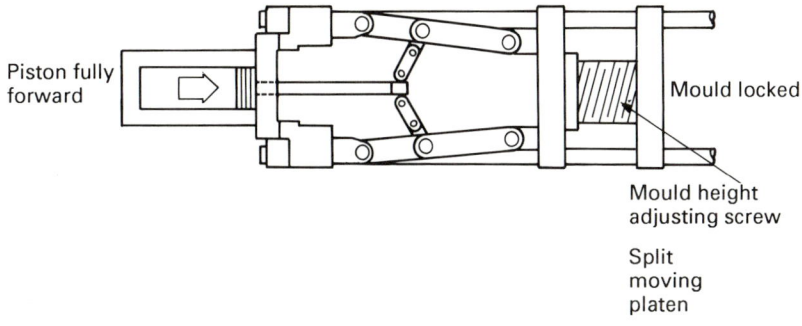

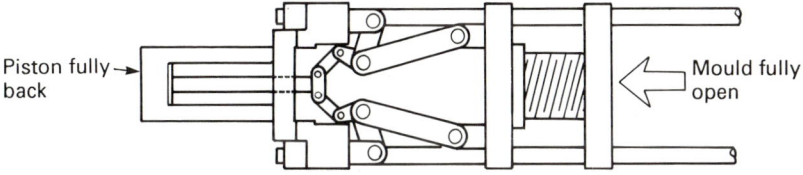

Figure 5.15 *Simple double-toggle mould locking system. Reproduced by courtesy of PPITB*

EXTRUSION

While there are fewer machines used for extrusion than for injection moulding it is probably true that, in tonnage terms, more polymer is subjected to extrusion than injection moulding. Not only is the process widely used to make finished products, such as film, piping, ducting, cable, hose and so on, but the process is also used for such intermediate operations as preparation of sheet for vacuum forming, mixing operations, and reworking of waste material. Furthermore the injection unit of an injection moulding machine and the corresponding part of a blow moulding machine are, in effect, captive extrusion operations.

In principle, extrusion consists of forcing fluid material through an orifice to give an *extrudate* of constant cross-section. The process may be applied to such diverse materials as pasta and metals, as well as to molten polymers. In the plastics industry the material is usually used in molten form and pumped to the orifice or *die* by means of a screw pump. The rest of this section will refer only to such screw extrusion processes.

SINGLE-SCREW EXTRUSION

Most extrusion operations use a single screw extruder. (Multi-screw systems will be mentioned briefly in a subsequent section.) A typical single screw machine is illustrated schematically in Figure 5.16. An extrusion line is made up of the following components:

1 material feed;
2 extruder barrel;
3 extrusion die;
4 haul off (and in some cases post-extrusion shaping).

Material feed arrangements must ensure that dry granules are fed to the hopper and that this is fed to the barrel of the extruder in a steady manner. For wire and paper coating operations similar comments apply to the wire and paper.

The function of the barrel and screw is to receive granules, plasticize them to the correct consistency and pump the materials into the die for shaping. There are a number of terms associated with screw design and these are shown in Figure 5.17. It is usual to consider the

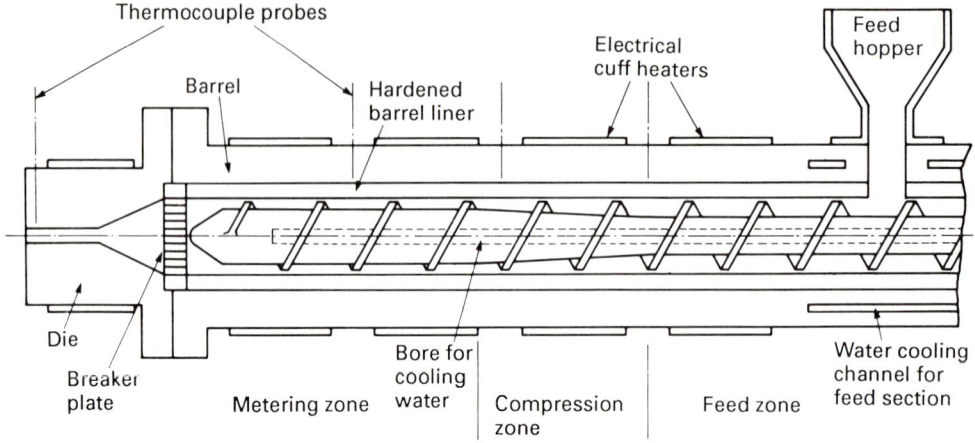

Figure 5.16 *The essential features of a single screw extruder. Courtesy Plastics and Rubber Institute*

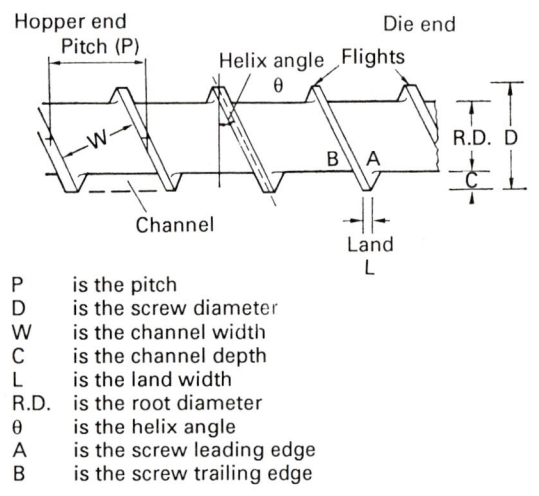

P   is the pitch
D   is the screw diameter
W   is the channel width
C   is the channel depth
L   is the land width
R.D. is the root diameter
θ   is the helix angle
A   is the screw leading edge
B   is the screw trailing edge

Figure 5.17 *Common terms associated with screw design. Courtesy Plastics and Rubber Institute*

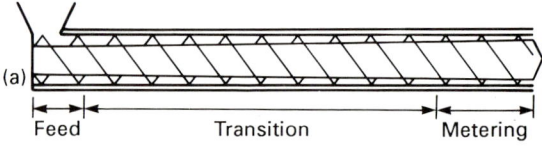

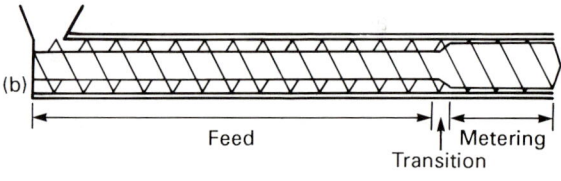

Figure 5.18 *The zones of an extruder barrel. In system (a) the transition zone comprises most of the barrel length whereas in (b) the transition zone is not more than one turn of the screw flight. Courtesy Plastics and Rubber Institute*

barrel in three zones each of which has a specific role. These are:

1 the feed zone;
2 the compression zone; and
3 the metering zone (see Figure 5.18).

The feed zone accepts granules from the hopper, pumps them up the barrel and commences heating. The channel depth must be adequate to take in the granules. One problem that may occur is that the granules fall into the screw channel, and adhere to the walls of the channel but not to the barrel wall. They then turn with the rotation of the screw without being pumped up the barrel. Efficient pumping

requires that the adhesion of granules to barrel is greater than adhesion of granules to the screw root. For many materials it is found that the polymer–metal adhesion is greatest near to the polymer melting point and more efficient pumping is obtained if the screw is cooled and the barrel is at a higher temperature. With other materials, UPVC for example, the difference in adhesion is less marked and it may be necessary in some instances to use a twin-screw extruder which has a more positive pumping action.

As the granules melt, the plastics mass fuses together trapping some air, and possibly gaseous degradation products. It is essential to force such gases out of the melt before it reaches the die. This is achieved by compressing the melt in the *compression zone*. In this zone the volume of one turn of the channel is reduced either by increasing the screw root or decreasing the pitch. For most thermoplastics it is more common to increase the screw root. The amount of compression is quantified by the *compression ratio* which is given by

$$\text{Compression ratio} = \frac{\text{Swept volume of one turn of channel at feed}}{\text{Swept volume of one turn of channel at exit}}$$

Compression ratios may range from 1.1:1 to 4:1 but a figure of the order of 2.25:1 is more typical.

Where the material softens or melts over a wide temperature range the compression zone may comprise a large portion of the total screw length. With sharp-melting materials, such as the nylons, the compression zone may be no more than one turn of the screw flight. Typical screw profiles for a number of materials are shown in Figure 5.19.

The function of the *melt zone* is to ensure that the material is in the correct molten state for feeding to the die. The quality of the melt may often be improved by increased shear made possible by reducing the effective channel depth in the feed zone. This effective channel depth may be reduced by screw cooling which freezes melt adjacent to the screw root and thus gives an additional operational variable. The efficiency of the pumping depends on the helix angle and is clearly zero both at helix angles of 0° and 90° being at an optimum at about 20°. In practice screws are commonly designed with a pitch equal to the screw diameter.

With some materials there is a tendency for gases to be evolved in the melt state. These gases may be due to moisture, to monomeric or other degradation products, or to volatile additives. These may be removed by the incorporation of a *decompression zone* in which the screw

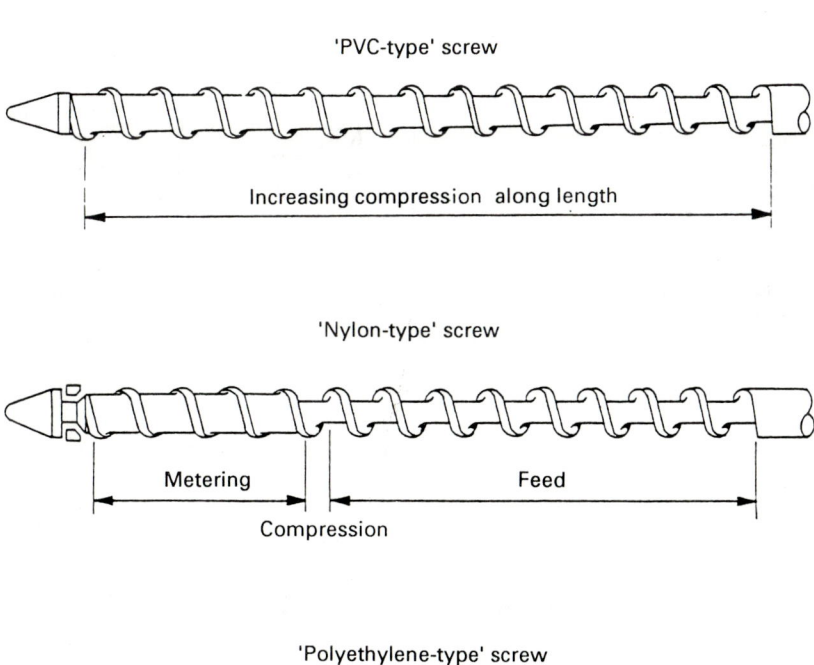

Figure 5.19 *Typical screw profiles for three types of thermoplastics*

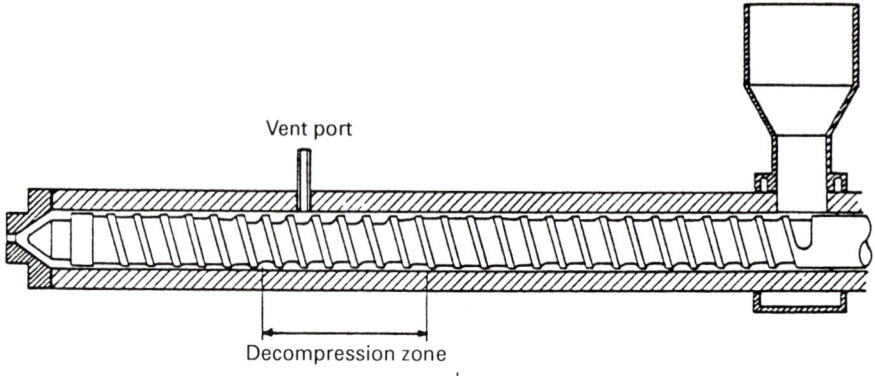

Figure 5.20 *Decompression-type screw used in conjunction with a vented barrel*

channel is increased. At this point, providing there is not an excess amount of restriction at the die head, the channel will not be full and gases will be able to escape through vents in the barrel wall (see Figure 5.20). Most extruders have screw length/diameter ratios (L/D ratios) of between 15:1 and 25:1, with lower values possible with the larger diameter screws. They may be somewhat longer with vented barrel systems containing a decompression zone. The barrel is divided into a number of heating zones (typically from 3 to 6) with a low temperature at the rear rising to a peak at the metering zone.

As melt emerges from the end of the screw it tends to move in a helical fashion. It is usually desirable to redirect the flow into an axial direction and this may be achieved by the use of a *breaker plate*, a perforated disc which not only aligns the flow axially with the barrel or die, but helps to build up a back pressure of melt in the barrel necessary to give good quality extrudates. In conjunction with this device a *screen pack* consisting of several layers of wire gauze may be used. This not only helps to build up the back pressure but will also hold back impurities and unplasticized material. As the screen packs tend to become clogged it is desirable to make provision for changing them without interrupting the extrusion process.

EXTRUSION DIES

On leaving the barrel the melt is fed into a die whose function is to determine the shape of the cross-section of the extruded product. Dies vary greatly in their shape according to the product

being made. Thus dies for rod, tube, sheet, film and wire covering will all be very different. They all, however, have the common feature of trying to ensure that all surfaces emerge from the die at a constant rate in order to avoid distortion of the product. The actual dimensions of the extrudate will, however, be affected by die swell (which increases the cross-section) and by the rate at which the extrudate is removed from the die (this is usually greater than the natural extrusion rate and so tends to decrease the cross-section).

Figure 5.21 shows a simple design for making a solid rod. This diagram shows the following important die features:

- D_D diameter of die orifice;
- D_B diameter of bore of extruder barrel;
- α lead-in angle;
- P length of *die parallel (die land)*.

The extrusion process is widely used for making tubing. A mandrel is fixed in the die to give a hollow extrudate. One such design is shown

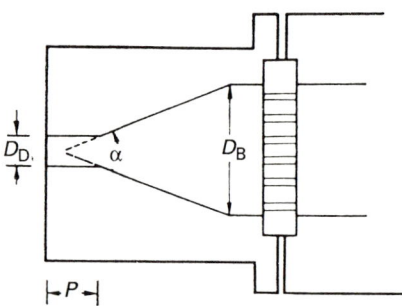

Figure 5.21 *Simple die design for making solid rod. Courtesy Plastics and Rubber Institute*

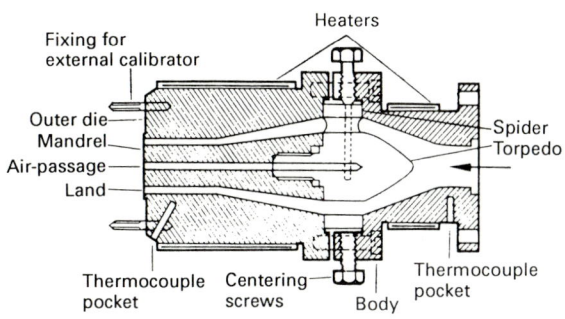

Figure 5.22 *Tubing die. Courtesy Plastics and Rubber Institute*

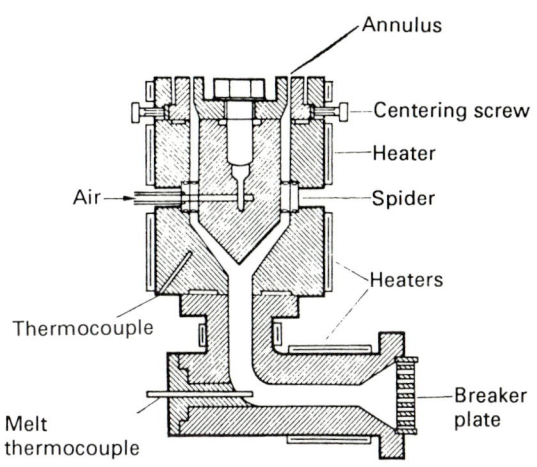

Figure 5.24 *Tubing die involving 90° turn in flow path via elbow adaptor. Courtesy Plastics and Rubber Institute*

in Figure 5.22. In this design the mandrel is located on a torpedo which is held in place by a spider. This normally takes the form of a ring, which is fixed into the die body with legs which connect to the torpedo, often via an inner ring. A simple spider design used to hold a mandrel for a garden hose is shown in Figure 5.23. The legs of the spider have an 'aerodynamic' cross-section to give smooth flow of melt. This should help the melt to re-fuse downstream of the spider and also reduce chances of material hold-up and stagnation.

Whereas the die shown in Figure 5.23 is of the *in-line* type (in which the extrudate emerges in the same direction as the extruder barrel axis) it is sometimes desirable for the extrudate to emerge at an angle. This may be achieved using an elbow adaptor as in the simple tubular film die shown in Figure 5.24. Figure 5.25 illustrates the use of a side-entry die, used in this case for wire coating. Figure 5.26 is a typical 'manifold' die design used for making sheet.

Film and sheet are usually made by *co-extrusion* processes. This may enable different surfaces to have different characteristics or to

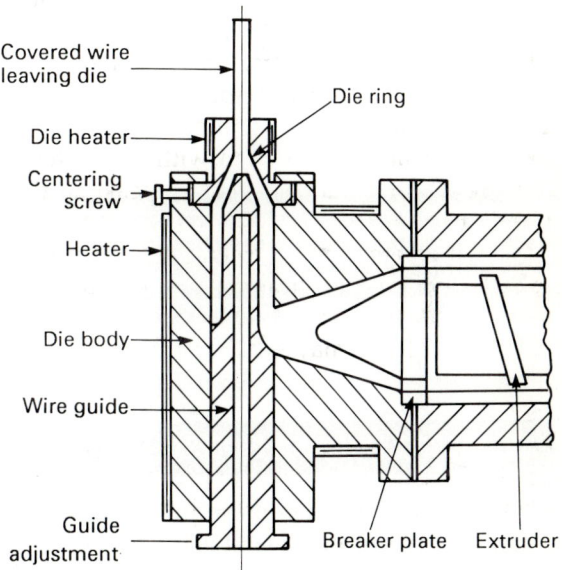

Figure 5.25 *Side-entry die. Courtesy Plastics and Rubber Institute*

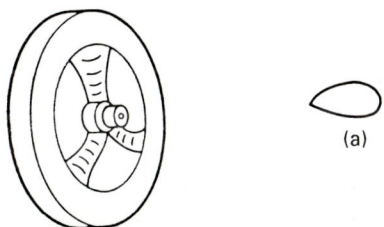

(a)

Figure 5.23 *Simple spider design showing (a) aerodynamic cross-section. Courtesy Plastics and Rubber Institute*

provide an inner layer with different properties to the surface. One such system is shown in Figure 5.27. In this system two melt streams from two extruders are fed to the die but kept separate until they reach the die lip. It is important to keep the melt channels concentric and the flow paths of similar lengths. Many co-extrudates now utilize more than two layers.

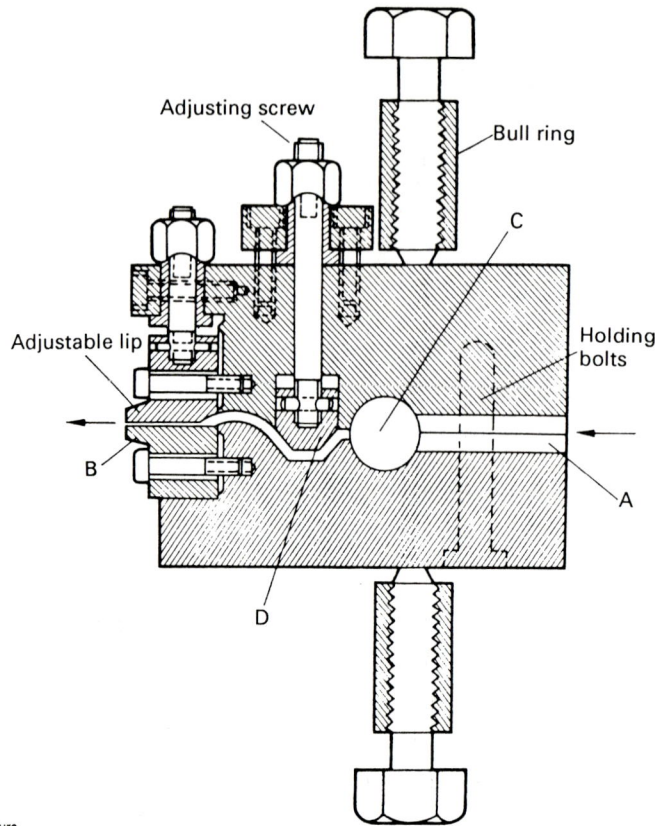

Figure 5.26   *Manifold die for sheet manufacture*

In this design, the melt passes to the die down an inlet tube (A), into tubular manifold (C) which is transverse to the extruder barrel axis. (In many designs the manifold may be slightly bent so that the distance between the manifold and the die lips is less at the edge of the die than at the centre in order to compensate for the pressure drop that occurs down the manifold.) Between the manifold and the die lips there is also an adjustable restrictor bar (D) which can 'fine tune' the flow rates across the width of the die. The melt then flows between a fixed die lip (B) and an adjustable die lip.

THE OVERALL EXTRUSION PROCESS

The extruder and die form just two, albeit essential, parts of the overall extrusion process. Provision must also be made to feed material to the die, to calibrate the extrudate to the correct size on emerging from the die, to cool the melt, to provide for a system for hauling the extrudate away from the die and for collecting the extrudate. With so many different products being made by extrusion processes the overall systems vary extensively. Some basic systems are shown below.

Figure 5.28 illustrates a system for solid and profile extrusion. An arrangement for hollow section extrusion is illustrated in Figure 5.29. Low density polyethylene blown film may be prepared using an arrangement similar to that illustrated in Figure 5.30. The principles of a cable covering layout are shown in Figure 5.31 and for extruded sheet in Figure 5.32.

MULTI-SCREW EXTRUDERS

In a single-screw extruder the forward pumping action caused by screw rotation (this is known as *drag flow*) is partially offset by a reverse or *pressure flow* caused by restrictions at the die head. Simplified extrusion theory states that for a single screw extruder the net output (Q) will be given by

$$Q = AN - B\Delta P/\mu,$$

where N is the screw speed, ΔP is the back

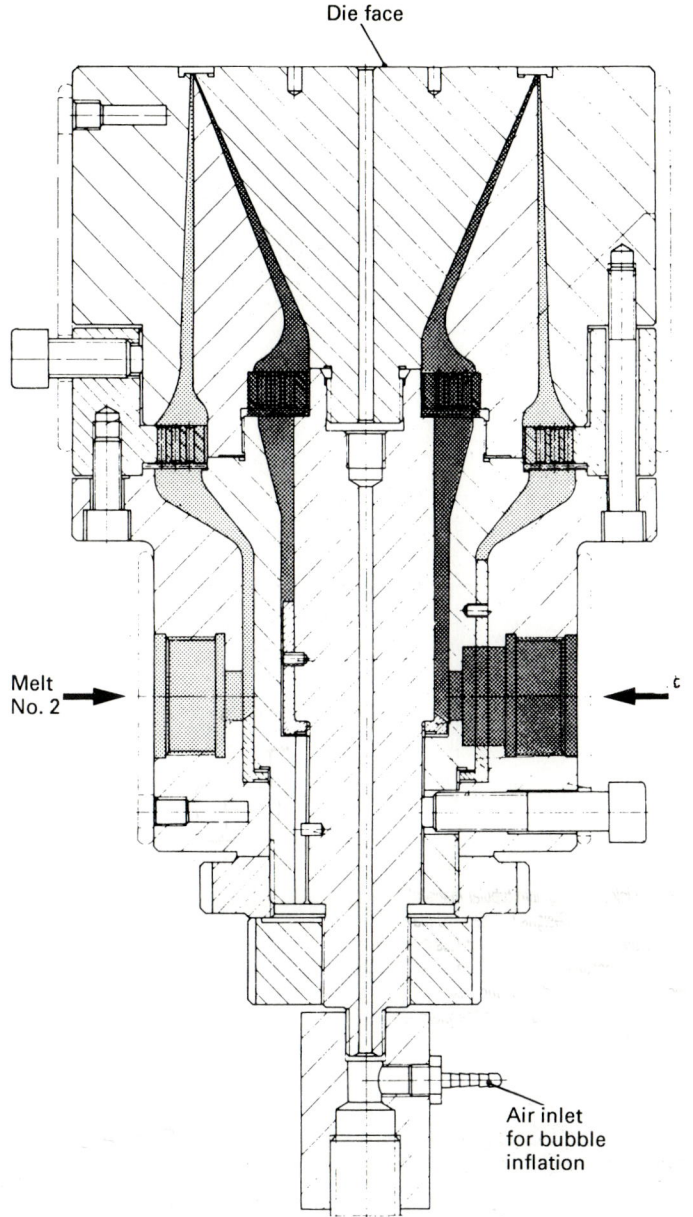

Die face

Melt
No. 2

Air inlet
for bubble
inflation

Figure 5.27   A coextrusion die for blown film. Courtesy of PPITB

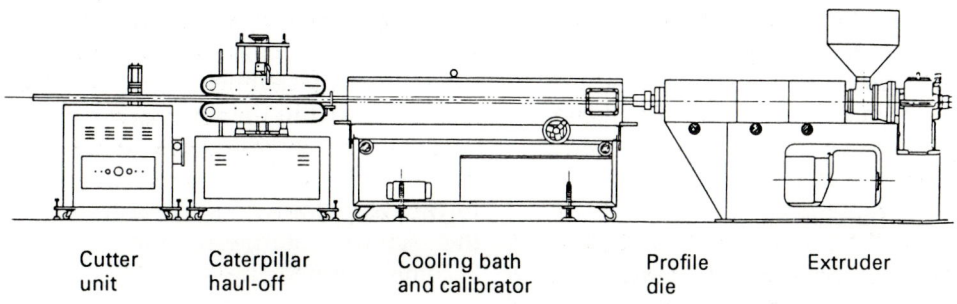

Cutter   Caterpillar   Cooling bath   Profile   Extruder
unit   haul-off   and calibrator   die

Figure 5.28   A typical line for solid and profile extrusion. Courtesy of PPITB

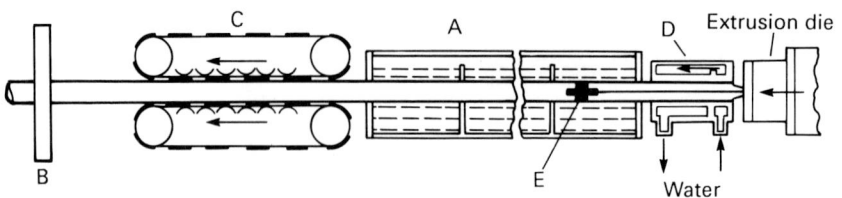

Figure 5.29 *Layout for manufacture of hollow tubing. A is the water tank; B is the cutting saw for cutting tubing into lengths; C is the caterpillar haul-off; D is the sizing die; E is the floating plug to prevent pressurizing air from escaping out of end of pipe. Courtesy Plastics and Rubber Institute*

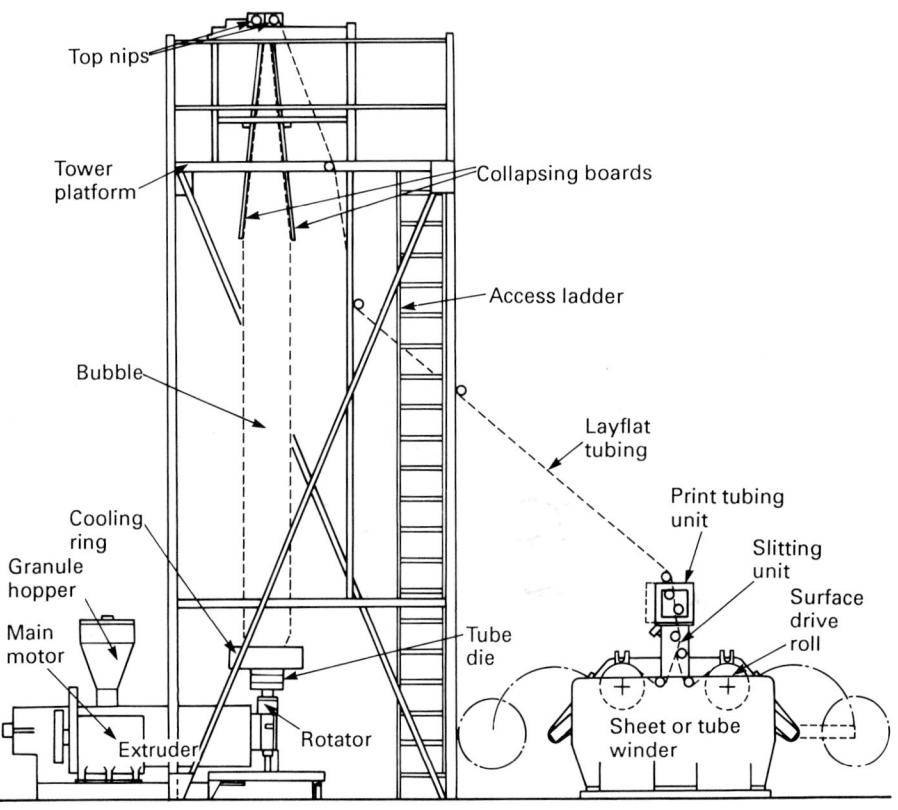

Figure 5.30 *The basic components of a blown film line for low density polyethylene*

pressure generated at the die head, μ is the melt viscosity and A and B are constants determined by the screw dimensions. (N.B. Whilst this formula is useful in explaining certain phenomena and also in providing approximate data it is based on a number of assumptions including the following

1 The melt is Newtonian (i.e. the viscosity is independent of shear rate).
2 The process is isothermal.

3 The channel depth and pitch are constant.
4 The channel depth is small compared to the width of the channel.
5 The channel in the melt zone is full.
6 There is no leakage across the screw flights.)

In the case of a twin-screw extruder, in which the screws intermesh, the polymer melts become trapped in small C-shaped spaces between the screw flights and are positively pumped up the barrel. In such circumstances

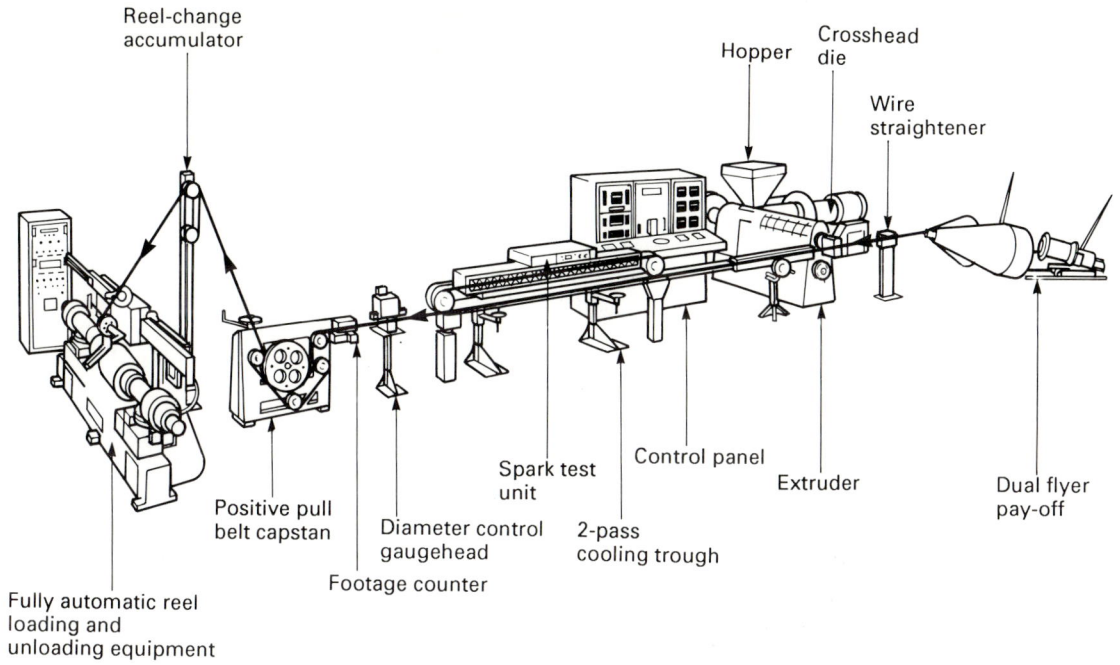

Figure 5.31   *A typical wire coating extrusion line*

Extruded Sheet

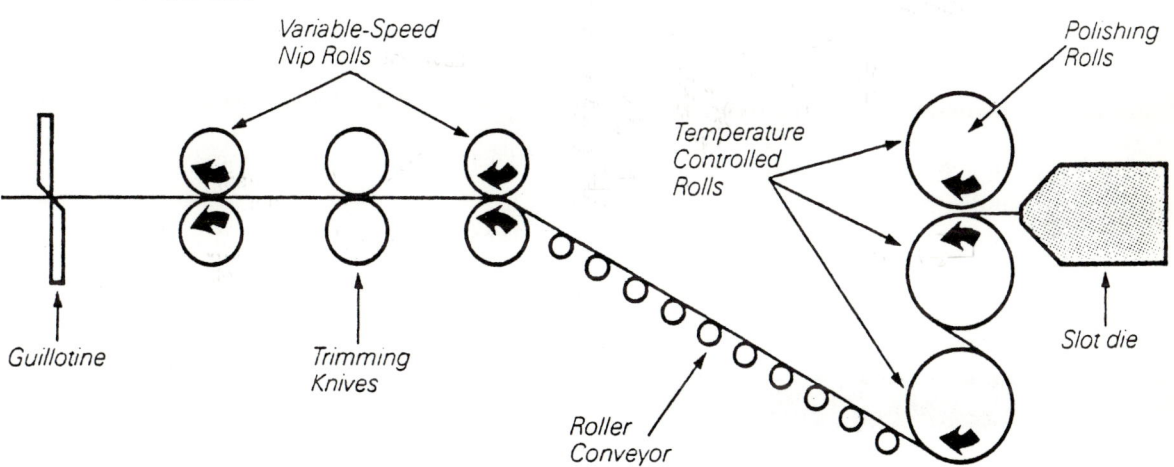

Figure 5.32   *An extruded sheet*

the output becomes independent of the die head pressure, i.e. the output can be expressed by the equation

$$Q = AN.$$

The two systems are compared in Figure 5.33. The screws of twin-screw extruders may be co-rotating or contra-rotating, the latter being more common (Figure 5.34). Extruders with

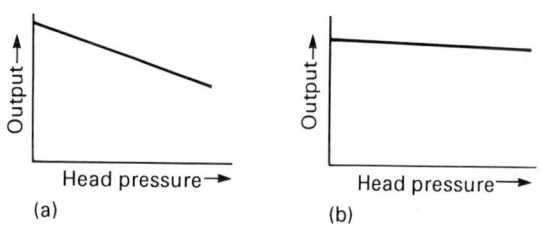

Figure 5.33   *Output – die head pressure relationships for (a) single screw extruder; (b) twin screw extruder with intermeshing screws. It follows that the output of such a twin screw extruder will be almost independent of the die orifice dimensions. Courtesy Plastics and Rubber Institute*

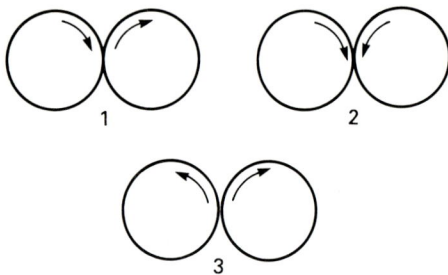

Figure 5.34 *Screw rotation in a twin-screw extruder. 1, co-rotating; 2 & 3, contra-rotating. Courtesy Plastics and Rubber Institute*

more than two screws have also been produced but are of little importance.

The main advantage of twin-screw extruders is that being positive in their action they are able to handle materials such as powders, pastes and flake feedstock and difficult materials such as unplasticized PVC. However, they tend to be more expensive, and limited to rather low extrusion speeds, die head pressures must be kept under strict control because of the limitations in the size of thrust bearings that are possible where there are two screws side by side, and the *bulk factor* of the feed stock should be close to the compression ratio. (The bulk factor is the ratio between the apparent density of the granules and the density of the solid product, and thus a measure of the extent to which the granules have to be compressed to consolidate them into a solid extrudate.)

BLOW MOULDING

In general use and specifically in this section the term blow moulding will be used to describe processes used to produce hollow products by inflation of a tube or *parison*. The processes of shaping sheet by use of air pressure are not considered under this heading.

Over the years very many variations of the blow moulding process have been devised. At the present time three main variants may be recognized:

1 conventional extrusion blow moulding;
2 injection blow moulding;
3 stretch blow moulding.

EXTRUSION BLOW MOULDING

The principles of extrusion blow moulding are shown in Figure 5.35. In the first stage a tube (known in this context as a *parison*) is extruded, usually downwards, between halves of an open mould. The mould is then closed around the parison which is then inflated by compressed air to the shape of the mould cavity. The formed shape is then allowed to cool until it is capable of being ejected from the mould and the cycle is repeated.

A great many variants in the process have been developed. These include:

1 Continuous or intermittent parison extrusion.
2 Variations in the number of *stations*.
3 Variations in the number of impressions.
4 Position of blow pins.
5 Methods of controlling parison dimensions.

In *continuous parison extrusion* the parison is extruded continuously and this process is generally used for the mass production of small thin walled containers of up to a gallon in size. In *intermittent parison extrusion* the parison is produced intermittently in a series of individual short lengths, the process being used for

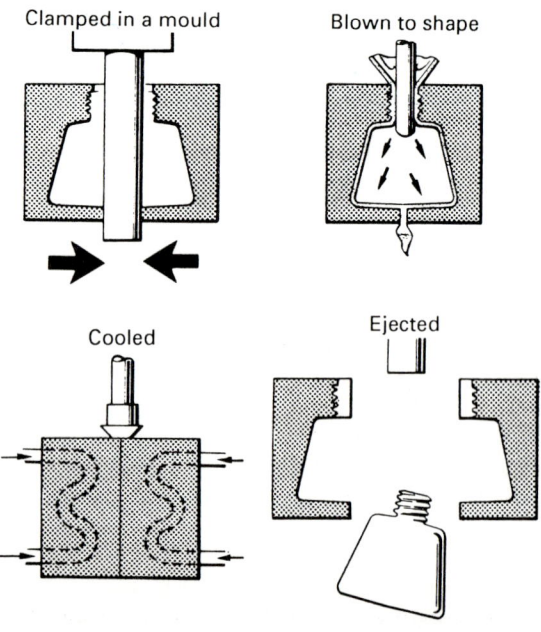

Figure 5.35 *The principles of extrusion blow moulding showing the four main stages of the process. Courtesy of PPITB. PPITB Standards Manual, Blow Moulding, P2–4.1*

larger thicker walled parts particularly for large containers and for industrial applications.

The continuous parison process may in turn be divided into

1 Single station systems – in which the machine has a single die head and one mould. The mould moves between the die and the blowing or calibrating station. Parison extrusion continues whilst the moulding is cooling in the mould (Figure 5.36).
2 Twin station systems – in which there is a single die head and two moulds which reciprocate between the two blowing stations and the die (Figure 5.37).

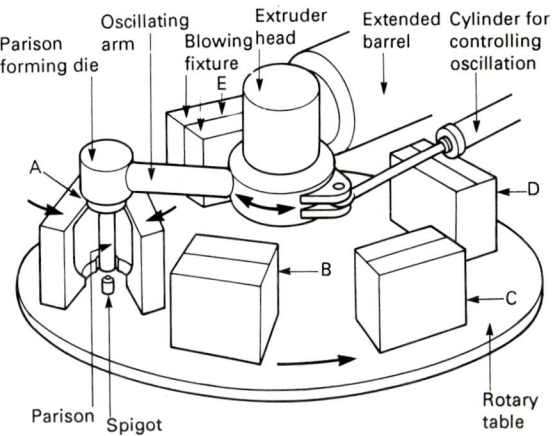

Figure 5.38 *Schematic arrangement for a rotary blow moulding system using continuous extrusion. Parison is being fed to mould A. At the same time inflation will be occurring in mould B and cooling in C, D and E.*

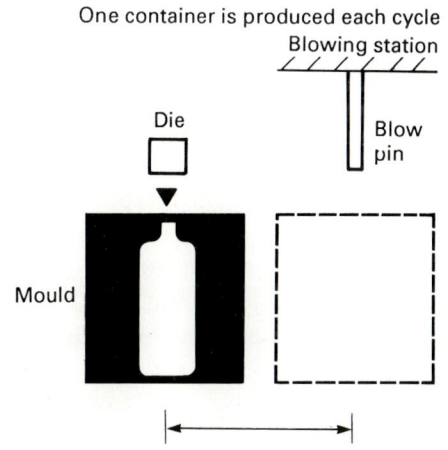

Figure 5.36 *Continuous parison extrusion blow moulding – single station system. From PPITB Standards Manual, Blow Moulding, p. P2–2.4*

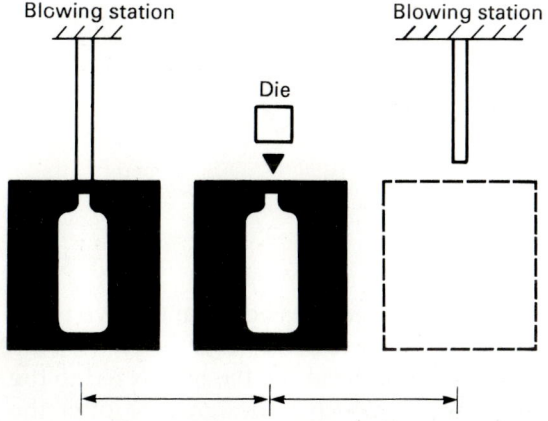

Figure 5.37 *Continuous parison extrusion blow moulding – twin station system. This arrangement has a single die head and two single cavity moulds which reciprocate between the die and its blowing station. Courtesy of PPITB. From PPITB Standards Manual, Blow Moulding, p. P2–2.4*

3 Rotary systems – in which the moulds are stationed on a rotary table (Figure 5.38).

In any of the above systems, and particularly with the reciprocating twin station systems, more than one parison may be extruded from a die simultaneously into a multi-impression mould. Figure 5.39 shows two four-cavity moulds producing eight containers per cycle. In the schematic system shown in Figure 5.35 the blowing was from the top (*top blowing*). In many processes, particularly with the intermittent processes, bottom blowing systems may be used. A third possibility is to inject air into the parison by means of a needle inserted into the side of the parison (Figure 5.40).

In the intermittent processes a specific predetermined length of parison is required. There are two general ways of achieving this:

1 The use of a reciprocating screw system. This is essentially identical to the system used in the in-line reciprocating screw injection moulding machines.
2 *Accumulator head systems* in which the melt is fed directly from the extruder to an accumulator chamber. When this has been charged with sufficient melt, the parison is formed by the melt being forced through the die by means of a hydraulically actuated piston. These systems are widely used (Figure 5.41).

Other combinations are possible, including the use of two extruders each feeding two die heads.

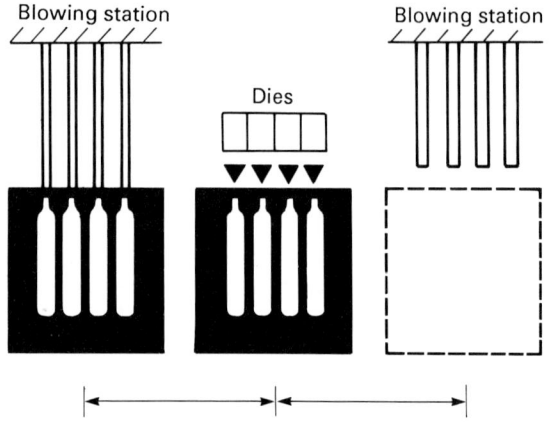

Figure 5.39 *As Figure 5.37, but in this case there are two four-cavity moulds giving eight mouldings per cycle. Courtesy of PPITB. From PPITB Standards Manual, Blow Moulding, p. P2–2.4*

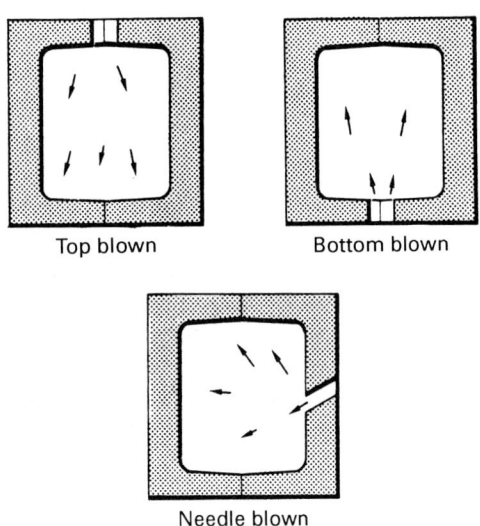

Figure 5.40 *Blowing geometries for extrusion blow moulding. From PPITB Standards Manual, Blow Moulding, p. B2–3.4*

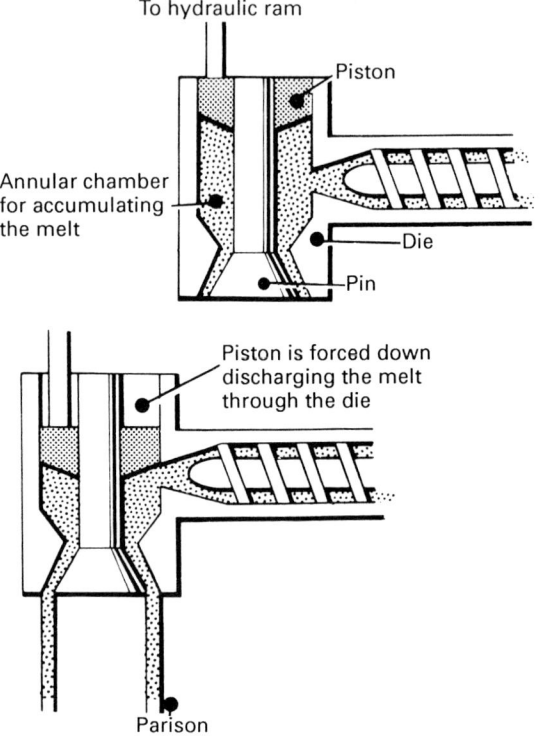

Figure 5.41 *Intermittent extrusion blow moulding – accumulator head systems. Such systems are mounted very close to the extruder and are widely used. From PPITB Standards Manual, Blow Moulding, p. B2–3.4*

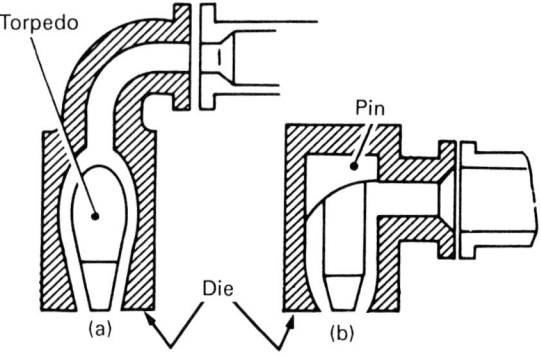

Figure 5.42 *Dies for extrusion blow moulding. (a) Torpedo-type. (b) Pin-head type. From PPITB Standards Manual, Blow Moulding, p. 6–1.1*

Somewhat similar are ram accumulator systems which use a quite separate ram and cylinder for melt accumulation.

Die heads and parison thickness control
Two main types of die are used for extrusion blow moulding:

1 torpedo head dies;
2 pin head dies.

The essential differences between the two types are shown in Figure 5.42.

In the torpedo head die the melt is fed to the top of the die which is designed to offer the minimum resistance to flow and minimization of possible *dead spots* where material may cease to flow (stagnate) and decompose. Torpedo head dies are particularly useful for heat sensitive materials such as PVC, polycarbonates and polyethylene terephthalate. In order to prevent the parison collapsing during extrusion, low pressure compressed air is usually supplied via the spider (not shown in the sketch) and the torpedo.

The side fed pin head dies are commonly used for extrusion blow moulding of polyethylene and polypropylene.

When a parison of uniform thickness is blown to shape in a mould, the moulding will vary in wall thickness from point to point according to the amount the parison had to stretch during blowing. Techniques have been developed to minimize this variation known as *parison wall thickness control*.

In the case of longitudinal variation, control may be achieved by the use of tapered dies and pins. By moving one with respect to the other the annular gap between die and pin, and hence the wall thickness of the parison may be varied (see Figure 5.43). Various types of parison control are available, including closed loop

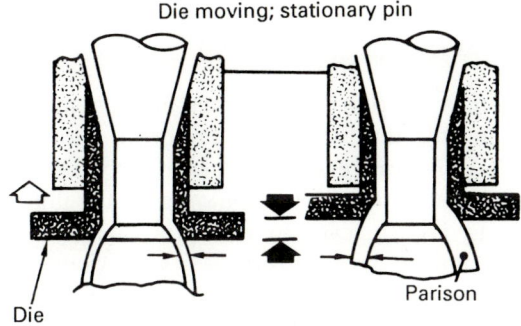

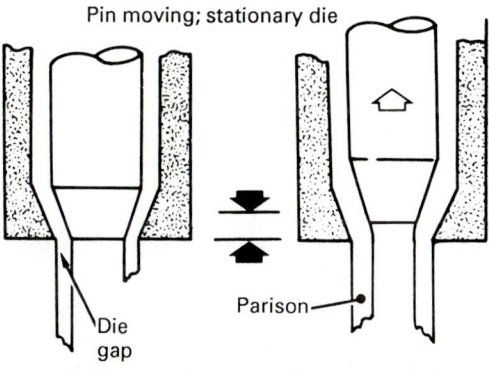

Figure 5.43 *Parison wall thickness control. The shape or cross-section of the parison can also be controlled by altering or adjusting the shape of the die gap. From PPITB Standards Manual, Blow Moulding, p. 6–6.1*

systems which monitor parison wall thickness, compare it with a programmed desired thickness and move die and pin relative to each other in order to achieve the correct thickness.

Control of wall thickness, circumferentially, may be achieved by the use of a flexible ring which may be deflected in and out around the circumference in order to match the desired thickness at any point.

INJECTION BLOW MOULDING

The process of injection blow moulding has been in existence for at least 30 years. For much of this time it was of somewhat lesser importance than the extrusion blow moulding process but recently a development of injection blow moulding (stretch blow moulding – described in the next section) has become of major importance.

The injection blow moulding process is shown in outline in Figure 5.44. In this process a parison is injection moulded directly onto a blow stick. The blow stick is then transferred, with the molten parison, to the blowing cavity. The parison is blown to the shape of the cavity by compressed air which is passed through the blowing stick.

In this process the weight of the parison may be accurately controlled, while the end of the parison may be properly formed rather than produced by squashing an extruded parison when the mould closes. The process is preferred where accurate control of weight and dimensions are of importance. One small disadvantage is that two parting lines may be observed on the moulding; the first from the parison mould and the second from the blowing mould. In the past, injection blow moulding has been used particularly for polystyrene containers for talcum powder and other toiletries.

STRETCH BLOW MOULDING

Stretch blow moulding processes have been introduced to enhance the level of crystallization in a moulded container. The following advantages may accrue compared to conventional extrusion blow moulding:

1 Lighter mouldings of comparable rigidity.

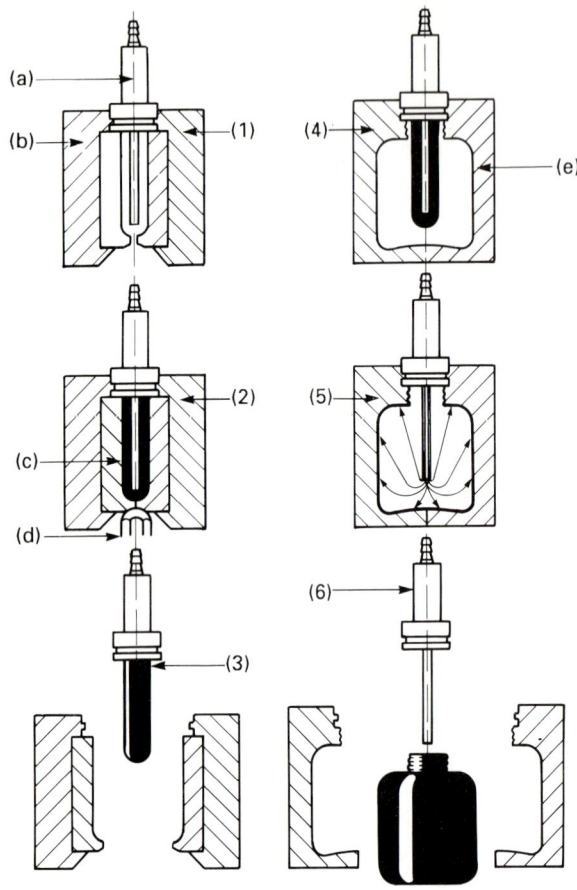

Figure 5.44 *Principles of injection blow moulding. 1, injection mould (b) containing transferable blowstick (a); 2, material injection via hot nozzle or runner (d) to form molten parison (c); 3, mould opens – blowstick removed with hot parison; 4, blowstick placed in blowing cavity (e); 5, air blown through holes in blowstick inflates parison; 6, blowing cavity opens – moulding removed.*

2 Mouldings have higher transparency and gloss.
3 Mouldings have reduced permeability.
4 Mouldings show greater precision and when made by injection stretch blow techniques are free from weld seams and bottom beads with increased burst resistance.

While the process may be used for PVC and polypropylene it has become of greatest importance for making PET (polyethylene terephthalate) bottles of particular use for beer and other carbonated drinks. One such process is illustrated schematically in Figure 5.45.

In the case of PET bottles the first stage is to produce an injection moulded preform *that is totally amorphous*. This will involve high melt temperatures of 270–280 °C (and hence thoroughly dried and highly pure high quality

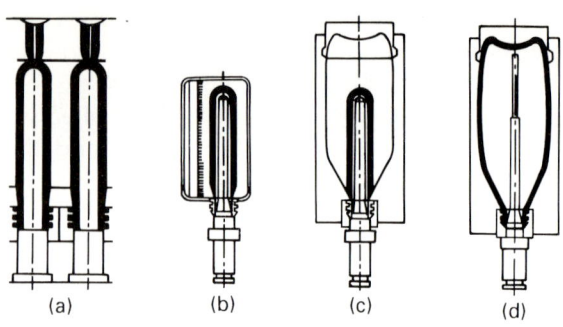

Figure 5.45 *Injection stretch-blow moulding process sequence. (a) Multiple injection moulding of amorphous preform. (b) Reheating stage. (c) Axial extension with telescopic mandrel. (d) Blow moulding.*

polymer), low injection pressures and low mould temperatures (between 4 and 20 °C).

The extracted preforms are then fed to a reheating device where they are raised to about 30 °C above the T_g, i.e. to about 100 °C. The reheated preforms are then immediately inserted into the blowing cavity where an air-activated telescopic mandrel stretches the preform longitudinally, while compressed air expands the preform to the shape of the mould cavity. In this way the polymer is stretched bi-axially, resulting in an increase in crystallinity.

Whereas the process described above usually uses an injection moulded preform, another stretch blow process (the Bekum process) is based on extrusion blow moulding and is illustrated in Figure 5.46. In this process the extruded parison is enclosed by a split preform mould in which the parison is first blown to the shape of the preform. The preformed shape is then transferred to the final blowing station in which, as in the previous process, longitudinal stretching is effected by means of a telescopic mandrel with simultaneous blowing to the shape of the cavity.

CALENDERING

The calendering process is a highly specialized one used for the manufacture of sheet and one which requires a high capital outlay. Originally developed for putting a sheen onto fabrics and paper, and for the manufacture of sheet rubber, it is used in the plastics industry for the manu-

In stretch extrusion blow-moulding the parison is

– clamped in a preform mould (this mould   – 'blown to shape'.
is smaller than the finishing mould)   – 'conditioned' or cooled to the thermo-
                                           elastic temperature of the material

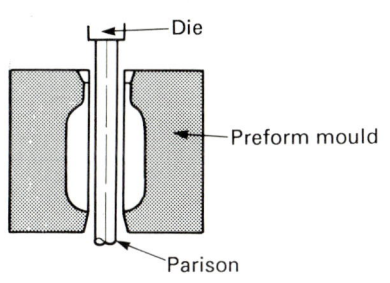
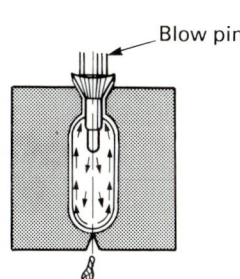

– transferred to and clamped in the   – stretched to shape by an extension of
finishing or stretch mould   the blow pin and by 'blowing air'

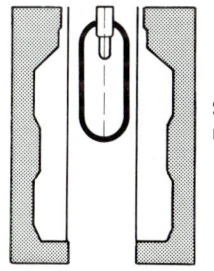
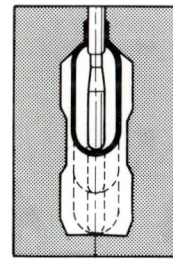
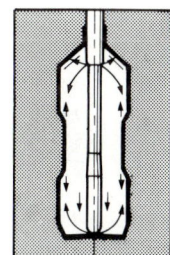

– cooled very 'rapidly'
– and ejected

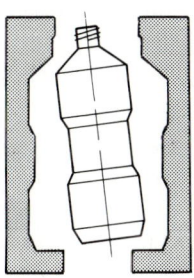

Figure 5.46   *Stretch blow moulding involving an extruder. Courtesy of PPITB. Copied from PPITB Standards Manual, Blow Moulding, pp. 2–4.1 & 2–4.2*

facture of PVC sheeting. With most other plastics materials extrusion processes are usually preferred because of the greater ease by which the latter are able to give products of high quality.

The calender consists of an arrangement (stack) of rolls (bowls) mounted in bearing blocks supported by side frames (gables) and equipped with roll drives, nip-adjusting gear, and feed, heating and haul-off arrangements. In outline the process consists of passing softened material between the rolls. For general PVC sheet production, four bowl machines are usually employed, with some use for three bowl machines in the manufacture of flooring. Some typical configurations are given in Figure 5.47.

The gap between a pair of rolls is known as a *nip* while the material just before the nip is known as the *bank*.

The high viscosity of the polymer sets up

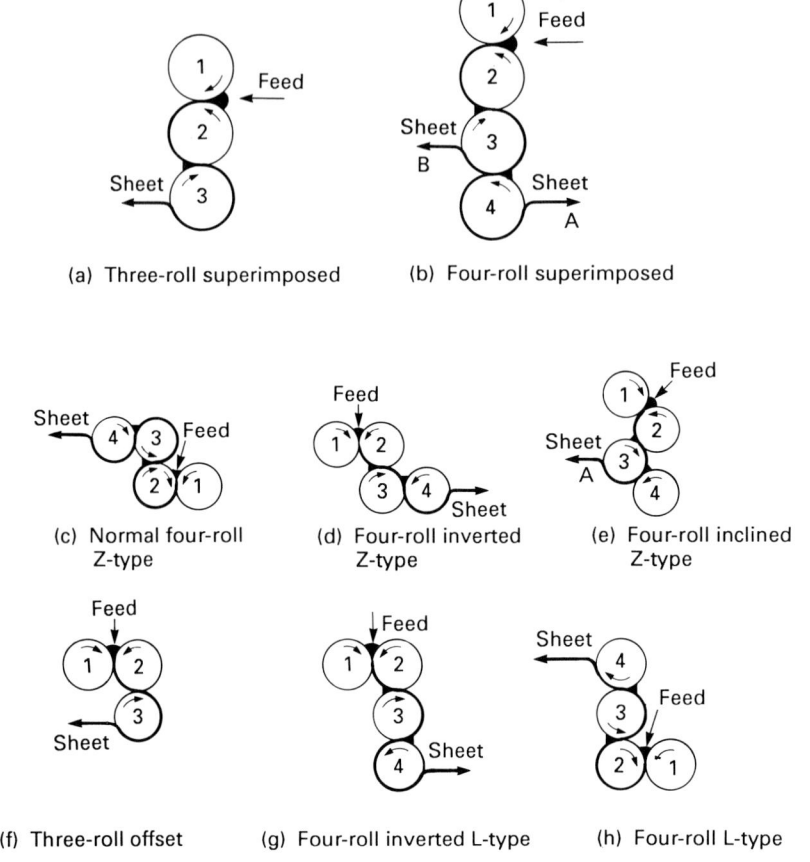

Figure 5.47 *Some common calendar configurations. Arrangements (c) and (d) differ only in the flow path of the plastics material through the calender. Courtesy: Plastics and Rubber Institute*

forces (which can be up to 1 MN per metre of working face) which tend to force the bowls apart and since they are constrained at their ends the bowls may bend slightly. Unless corrected this will cause the sheet to be thicker at the centre than at the edge. This can be corrected by:

1 Crowning the rolls (a procedure suitable only for a narrow range of conditions).
2 Roll bending. Achieved by applying a hydraulic load to the journal ends which exerts leverage on the roll against the bearings and in the absence of polymer would cause the bowl to be slightly concave at the nip. It may thus balance out the roll bending.
3 Roll-crossing (roll skewing, cross axis roll adjustment). In this case the rolls are slightly closer at the centre than at the edges. With both roll crossing and roll bending the com-

pensating profile does not exactly match the original bowl bending due to the polymer.

The calender itself is only a small part of the overall calendering process. The calender line commences with mixing stages, metal detection and prewarming stages, with material usually fed in strip form to the calender. After calendering there are several stages and a typical post-calender section layout is shown in Figure 5.48.

In this design an L-type calender is used, the sheet being removed and passed round a series of multiple stripping rolls. Where it is desired to emboss one of the surfaces the sheet is then rewarmed, usually by infra-red heaters, and passed between a chilled metal roller engraved with the emboss pattern and a larger diameter synthetic rubber-covered backing roller. The sheet is subsequently passed round a series of cooling rollers, through a β-ray thickness gauge,

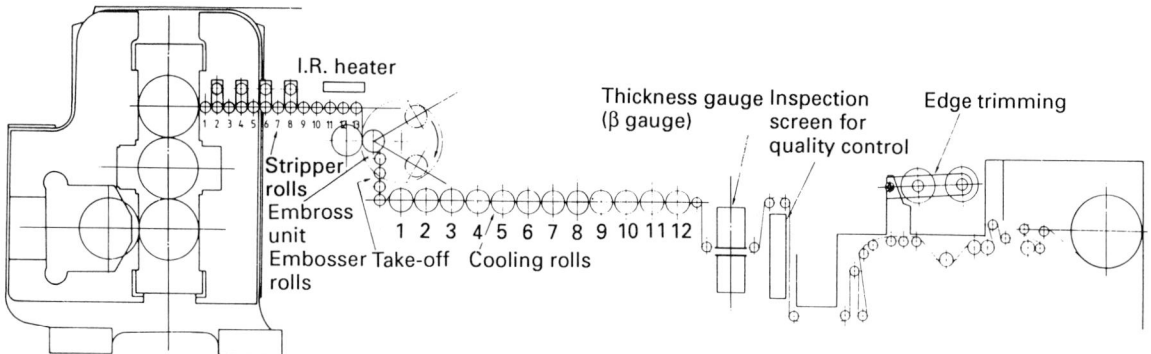

Figure 5.48 *An 'L' calender with post-calender section. Schematic representation. After Titow, PVC Technology (4th ed.), Applied Science*

past an inspection screen, an edge trimmer and on to a wind up. Great care must be taken not to unevenly stretch the sheet during these stages.

THERMOFORMING

The term *thermoforming* is somewhat misleading since it implies shaping using heat, a feature of most plastics processing operations. In practice it is used for processes in which the polymer is warmed to the *rubbery* rather than the *molten* state and shaped in that state. After shaping the shape is then retained simply by cooling. Subsequent reheating will cause the shaped part to revert to its pre-shaping dimensions and it is best to think of thermoformed products as frozen distorted rubbers. In some cases the thermoforming operation consists of stamping or coining prewarmed solid blanks in which case the process is sometimes referred to as *solid state moulding*. This section is however confined to a brief discussion of shaping of sheet, this process being known as *sheet thermoforming* (and often simply as thermoforming).

Sheet thermoforming can be divided into three categories:

1 mechanical shaping;
2 vacuum forming methods;
3 compressed air techniques.

MECHANICAL SHAPING METHODS

In principle the simplest shaping operation involves pressing the softened sheet between the two halves of a matched metal mould (Figure 5.49). Such a process is of little value since the manufacture of two mould surfaces is unnecessarily expensive while furthermore the sheet will be marked on both surfaces. In some cases the desired shape may be made by pushing a simple skeleton jig into one side of the sheet and holding in position until the sheet has cooled (Figure 5.50). Mechanical shaping is however of limited utility and better results may usually be obtained by the use of differential air pressures on the two sides of the sheet in order to obtain the desired shape.

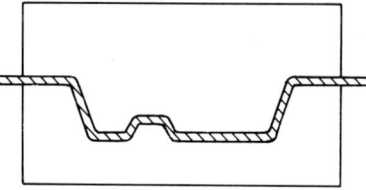

Figure 5.49 *Thermoforming using matched metal moulding*

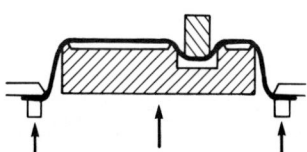

Figure 5.50 *Thermoforming using skeleton tooling to reduce contact of sheet with tool and hence reduce marking*

VACUUM FORMING METHODS

In the basic vacuum forming process, thermoplastic sheet is clamped over the mould, heated to the rubbery state and air is evacuated from the space between the sheet and the mould by application of a vacuum. The sheet is then, in effect, pushed onto the mould by means of the atmospheric air pressure which at sea level is about 0.1 MPa (14.5 p.s.i.). This then is the maximum pressure available for shaping and will be less when the vacuum is partial and at elevated altitudes. Thus without some form of assistance by compressed air or by the use of plugs and rams, vacuum forming is restricted to polymers of quite low modulus in the shaping range. For similar reasons it is also more difficult to shape thick sheet than thin sheet. Many variations of the vacuum forming process are known of which the following may be considered as the most important:

1 Female moulding (Figure 5.51). The heated sheet is sucked into a female mould cavity. Greatest thinning occurs at the base of the cavity with marking of sheet on the lower side. It is not difficult to make multi-cavity mouldings (or a single moulding with many cavities) using this technique but thinning may be excessive.

2 Male moulding (Figure 5.52). The sheet is drawn down over a male mould in the mould box. Since the sheet touches the top of the male former first and cools this will be the thickest part of the moulding (i.e. opposite to that obtained using female moulding). Marking will also be on the opposite side to that with female moulding. One difficulty that may arise, is that where the moulding has several male shapes, the sheet may form webs between the shapes. Increasing the separation between these shapes and/or lowering their elevation will help to reduce webbing.

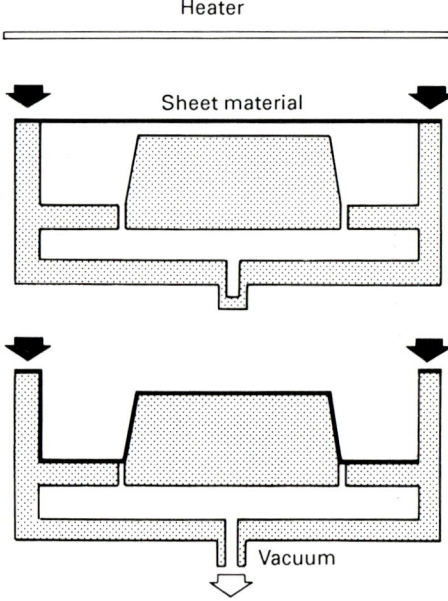

Figure 5.52   Vacuum forming – male moulding. From ICI Technical Service Note, G109 (4th ed.)

3 Drape forming (Figure 5.53). In practice, in the male moulding process shown in Figure 5.52, there will be extensive stretching of the sheet much of which will be wasted as trim. It is more economical to use the drape forming process in which the male former is first pushed up into the sheet by means of a ram followed by application of the vacuum. In this case the sheet does not have to be stretched so much.

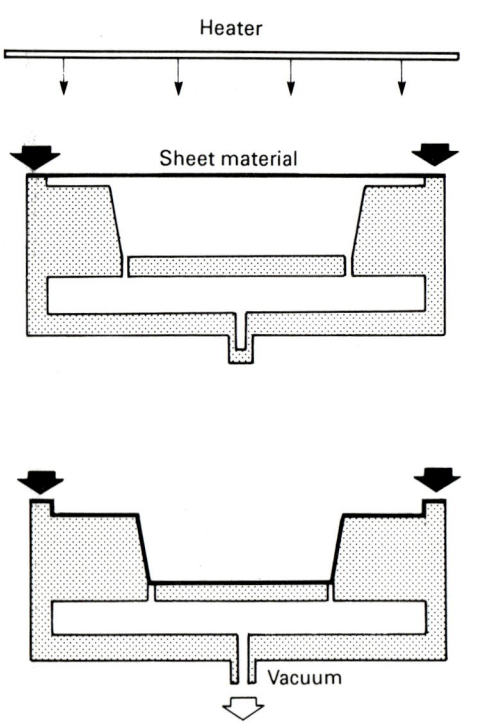

Figure 5.51   Vacuum forming – female moulding. From ICI Technical Service Note, G109 (4th ed.)

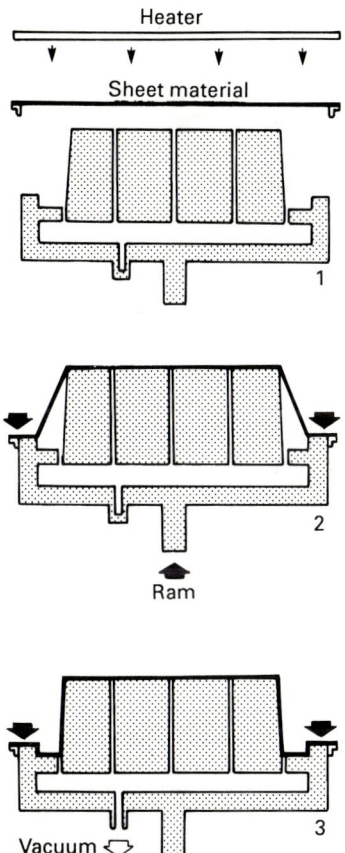

Figure 5.53 *Vacuum forming – drape forming over a male mould. From ICI Technical Service Note, G109 (4th ed.)*

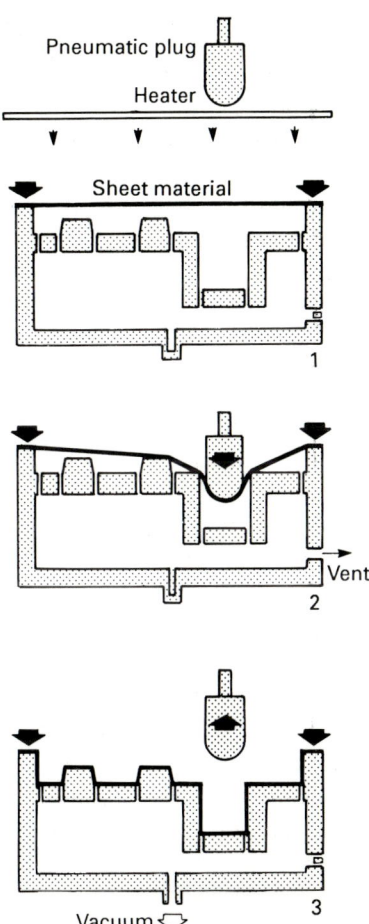

Figure 5.54 *Vacuum forming – using plug assist. From ICI Technical Service Note, G109 (4th ed.)*

4 Plug assist (Figure 5.54). Plug assistance methods are particularly useful with female moulding. In this case a plug, with a slightly smaller area than the female cavity, is lowered onto the heated sheet just before vacuum is applied. This chills the sheet over the area which would normally be subject to the greatest thinning in female moulding and thus helps to even out the stretching. Marking occurs on both sides of the sheet although on the plug side this may be minimized by using a skeleton as opposed to a solid plug. The process is not satisfactory with thin sheet because of difficulties in fully shaping the chilled sheet.

Somewhat better thickness distribution can sometimes be achieved if the sheet is pre-stretched by blowing into a shallow dome by means of compressed air fed to the space between the sheet and the mould just before the plug contacts the sheet. This pro-cess requires good control of the bubble height and the timing of the plug movement.

5 Drape forming with bubble (airslip or pre-stretch) (Figure 5.55). In this modification of the male drape process, the sheet is first inflated into a bubble in which the greatest thinning will be at its apex. The male mould is then raised into the bubble and vacuum applied. In this second stage the greatest thinning will be at the side of the bubble. This technique will normally give a more even thickness distribution than the other processes.

Further variations of vacuum forming include:

1 Vacuum covering and laminating. The sheet is vacuum formed onto a substrate, a method used particularly for automobile interior trim.

103

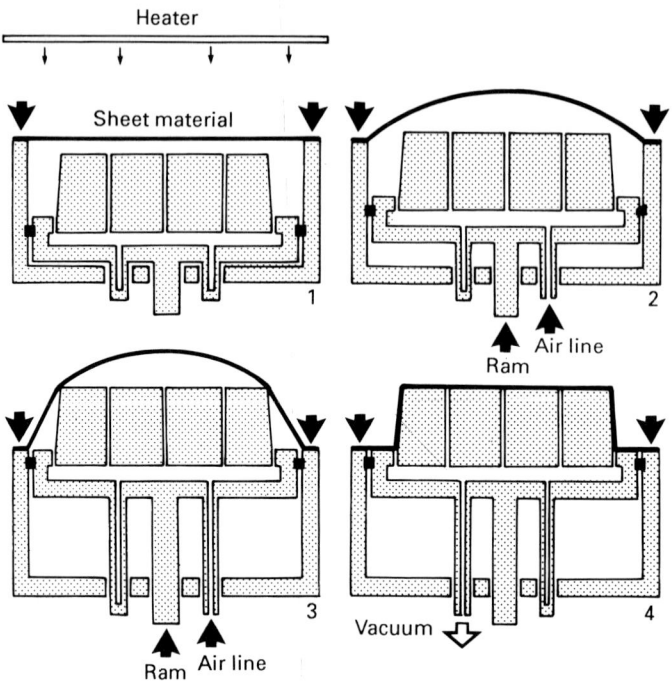

Figure 5.55 *Vacuum forming – air-slip, or bubble, forming.*
From ICI Technical Service Note G109 (4th ed.)

2 Skin packaging. The parts to be packed are laid on printed perforated card which has been coated with a heat activated adhesive. Thermoplastic sheet is drawn down onto the card by vacuum (air escaping through the perforations in the card) and as the material draws tightly around the product the adhesive is activated, bonding the formed film onto the card.

3 Blister packaging. The components to be packed are dropped into a preformed blister. Printed card coated with a heat activated adhesive is then pressed onto the blister using a heated platen. In blister packaging the packaged part is loose in the blister.

4 Twin-sheet forming. Two female moulds are used in which one is inverted over the other. A layer of sheet is clamped onto or nearby each mould box and heated. Vacuum is applied drawing the sheet into each of the two mould cavities which are then closed against each other, allowing the two moulded sheets to bond at the edges. This technique has been used for hollow parts such as for wind surfing boards.

Vacuum forming equipment may be of the standard 'off the shelf' type or individually built to meet customers' specific requirements. Most standard machines have facilities enabling them to use most of the techniques described above. Standard machines range in forming area from about 0.25 m × 0.25 m to 4 m × 2 m. Some machines are designed to handle separate pieces of sheet, others are fed by continuous lengths of sheet or film from a reel. In most machines movement of moulds, plugs, heating tables and sheet clamps is by pneumatic cylinders, while hydraulic systems may be used where high forces are involved. Mechanical clamping may be used on small machines.

In vacuum forming, sheet is usually softened *in situ*, i.e. above or adjacent to the mould as opposed to being heated in a convection oven. Conduction (contact) heating may be used for blister and skin packaging, but for general purposes, radiation heating using energy in the near infra-red region with a peak output at a wavelength of about 3 microns, is the fastest and most economical method.

Because of the low shaping forces involved, a wide variety of materials may be used for making vacuum forming moulds. These include:

1 Plaster. Plaster moulds are cheap to make but have a short life tending to crack on heating. Some artificial stone materials may be cast in the same way as plaster but have greater durability.
2 Wood. Best results are obtained with resin free fine-grained woods but it is very difficult to prevent grain marks from appearing in the finished moulding.
3 Thermosetting laminated materials are more expensive than wood but give a better finish and last longer.
4 Epoxide (epoxy) resins may be used to give inexpensive moulds but care has to be taken to ensure that they do not overheat during continual forming operations.
5 Metal moulds. Many types of metal mould may be used including sprayed metal moulds, cast metal moulds (usually aluminium alloy) and machined metal moulds. Metal moulds (particularly the cast and machined types) are very durable and particularly suited to long production runs.

AIR PRESSURE FORMING TECHNIQUES

Where atmospheric pressure alone is not sufficient to shape a sheet of thermoplastic material, and the use of plug assist techniques is not appropriate, the use of positive air pressure techniques may be called for. In such techniques air pressures up to 0.7 MPa (*c.* 100 p.s.i.) may be used.

A very simple example of the technique is that of *free blowing* in which heated sheet is clamped by a ring over a compressed air source on a blowing table. Application of the compressed air will cause the sheet to blow to the shape of a dome which will be free from any marking other than from the clamps. Some variation may be obtained by varying the shape of the clamping ring and by the use of skeleton tooling.

More generally, sheet may be clamped over a mould box as in vacuum forming operations, but rather than the sheet being sucked onto the mould a pressure box is clamped over the sheet

after it has been heated which allows compressed air to blow the sheet onto the mould.

Pressure shaping techniques are useful for

1 thick sheet materials;
2 materials of high modulus;
3 materials with low hot strength (which because of the higher pressures involved may be shaped at lower forming temperatures).

Because of the higher pressures involved, moulds and other equipment are generally more robust than is necessary for vacuum forming.

REACTION MOULDING

The term *reaction moulding* is used to describe processes in which chemical reactions occur during the mould process. In common usage the term has been largely restricted to reactions involving polyurethanes and polyamides. In the view of the author this is unduly restrictive and in this chapter the term will also be taken to include compression moulding of thermosetting plastics and the manufacture of fibre-reinforced thermosetting laminates.

COMPRESSION AND TRANSFER MOULDING OF THERMOSETTING MOULDING POWDERS

Compression moulding is the oldest method of heat shaping plastics materials. In the case of thermoplastics, the process is lengthy since heating and cooling of the mould are necessary in each cycle. More important is its use with thermosetting plastics, both of laminates and of moulding powders. This section will be confined to a discussion of the latter and to the related transfer moulding process.

The compression moulding process for thermosetting moulding powders is illustrated in outline in Figure 5.56. Slightly more material than required to make the moulding is charged into the cavity of a heated mould. The mould is then closed under pressure (usually in a hydraulic press). Under the influence of heat, the moulding material melts, and under pressure it

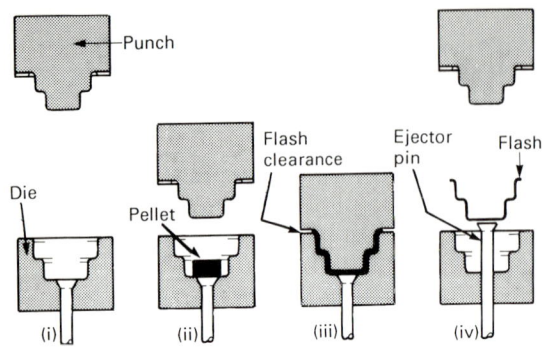

Figure 5.56 *The moulding sequence in compression moulding. (i) Mould open, cavity cleaned; (ii) pellet loaded, mould closing; (iii) mould closed – curing stage; (iv) mould open, moulding ejected. In the illustration a pellet (possibly preheated) is used but frequently, and particularly with automatic processes, powder is used*

flows to the shape of the mould cavity, provision being made for excess material to be squeezed out of the cavity as *flash*. On continued heating in the cavity the moulding material cross-links or *sets* and when the moulding is sufficiently hardened the mould may be opened and the moulding removed. The cross-linking process is commonly known as *curing*.

In a commercial operation the process may have four stages:

1 pelleting (an optional stage);
2 preheating (also optional);
3 the moulding stage;
4 finishing operations.

Pelleting permits accurate measurement of the charge to the mould, reduces contamination and facilitates preheating. It does not, however, always fit in very well with more automated compression systems.

Preheating is useful in that it reduces expensive moulding time, allows rapid heating of large pellets or masses of powder, helps to remove moisture and other volatiles prior to moulding and because it advances the cure has been claimed to reduce moulding shrinkage.

The optimum *cure* (i.e. cross-linking) for one property is not necessarily the best cure for another. It is therefore necessary to establish what properties are important in the finished moulding and use cure times and temperatures that give a good compromise to the various

requirements, including that of cost. It is particularly important that flow and cross-linking processes are controlled and are in the correct sequence. While improvements in the operation can often be made by such refinements as *breathing (venting)* they will be of no value if the basic process is not correct. A typical cycle is shown in Figure 5.57.

Since the material will start to cure as soon as it comes into contact with the heated mould it is important that flow and shaping be complete before the *gel* point is reached (i.e. the material shows first signs of being cross-linked). Deformation of partially cured resin may lead to undue stresses in the moulding; in severe cases flow may not be sufficient to give a properly formed moulding.

Having selected the general type of moulding material, the moulder will have to pay attention to the following material variables:

1 flow characteristics;
2 curing characteristics;
3 particle characteristics ('grinds').

There are three independent flow properties of concern in thermoset moulding:

1 The ease of flow (fluidity).
2 The total flow that occurs before the material sets.
3 The time available for flow.

Fluidity is the reciprocal of viscosity and at constant temperature does not change very much up to the time that the material gels. Thus the total flow that occurs will be related to the product of flow time and fluidity.

Moulding powders may be classified as *soft flow (easy flow, free flow)* at one extreme and *stiff flow (hard flow)* at the other. Free flow powders tend to be used where extensive flow, as in deep draw mouldings, is required or where there are delicate inserts. There is, however, a problem that in some moulds, free flow materials may flash prematurely leaving 'short' regions in the moulding. Stiff flow powders (in which the resins are of higher molecular weight) often give a better finish and it is often considered that the best practice is to use the stiffest powder suitable for a given mould.

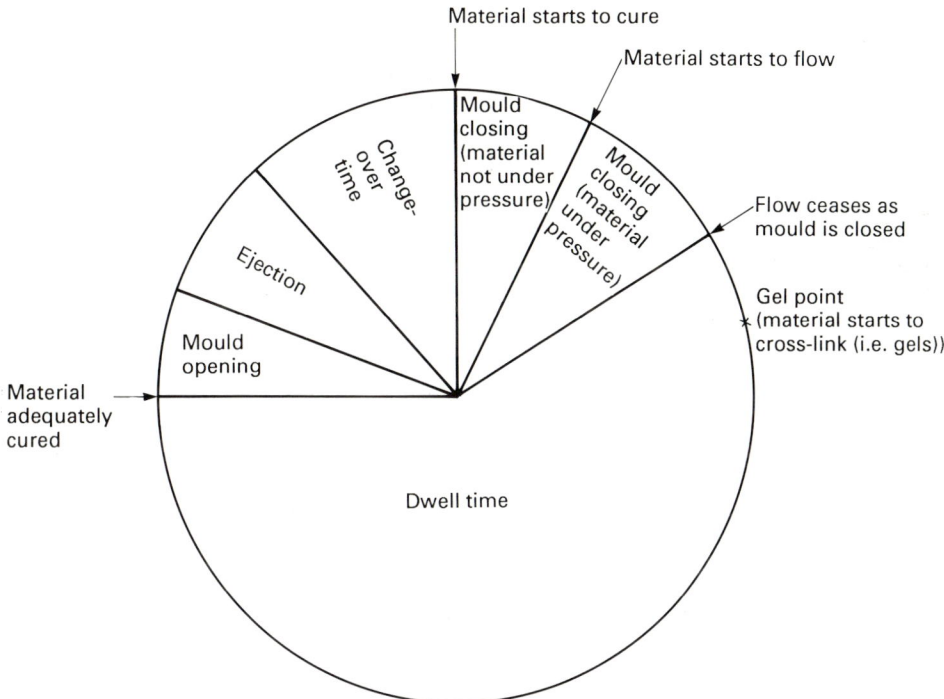

Figure 5.57 *The compression moulding cycle. In many instances the dwell time will comprise a larger fraction of the moulding cycle than indicated here. (It is instructive to compare this diagram with that for the injection moulding cycle, Figure 5.3)*

The moulding powders may also be supplied with a variety of cure rates. Where little flow is required, such as in moulding a large flat surface, a fast curing powder is desirable, but where extensive flow is needed then a slower curing grade may be used.

Manufacturers may also supply material in a range of grinds. Fine powders give the best finish but do cause dust problems. Conversely coarse powders are less dusty but give a coarse grained and often 'orange peel' surface. (This problem is usually less serious if extensive flow is involved.) For automatic moulding, free flowing granules are required, and for this purpose regular particles of a narrow particle size distribution are usually preferred. On the other hand, for pelleting, a wider spread of particle sizes including some *fines* is desirable to aid pellet consolidation.

Transfer moulding may be considered as a process intermediate to compression and injection moulding. The process involves heating the material in a transfer pot and when it is in a

suitable state of flow pushing it through a runner and gate into the mould cavity in which the material sets. In its simplest form the pot and cavity are all part of the same mould. Such an *integral pot mould* is illustrated in Figure 5.58. In more advanced designs separate pot systems may be used (and thus interchangeable with different cavities) and in some cases become almost indistinguishable from a ram injection moulding system.

Transfer moulding has been useful for moulding thick section products and also mouldings with thin sections and delicate inserts. This is because the material is usually in a better state of flow when it enters the mould cavity than is the case in the initial stage of compression moulding where semi-hard granules are being somewhat crudely forced around the mould cavity.

With the development of injection moulding of thermosetting materials it has become (as also has compression moulding) of less importance than hitherto.

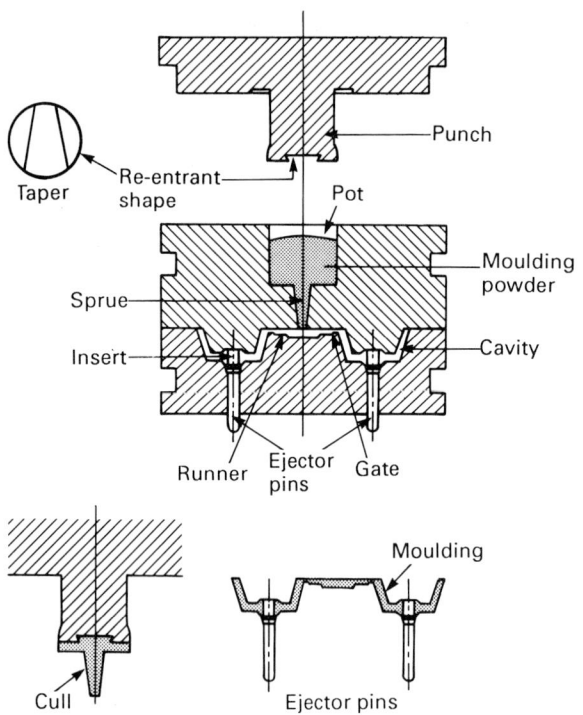

Figure 5.58 *Transfer mould of the integral pot type. More complex designs with separate pots approach injection moulding in their concept*

FIBRE-REINFORCED THERMOSETTING COMPOSITES

This section will deal only in briefest outline with manufacture of thermosetting laminates. A broader introductory review to the subject of composites in general is given in Chapter 1.

Laminates from phenolic and aminoplastic resins have been available for well over half a century. They are made by impregnating the reinforcing material (usually paper or cloth) with resin (usually in solution form), plying the layers of impregnated material and pressing them under high pressure. The high pressure is required to prevent gases evolved during cure from creating bubbles and other moulding defects.

In the 1940s resins became available which did not give off volatiles during cure, and moulding was possible at very low pressures. The appearance of the polyester laminating resins and later epoxide resins also allow curing to take place, if required, at room temperatures. This enabled very large mouldings to be made by *hand lay-up* techniques on simple moulds.

In the hand lay-up process a mould release is applied to the mould surface. Glass reinforcement, most commonly in the form of chopped strand mat, is then laid onto the mould surface and impregnated with liquid resin containing a curing system, often with a brush. Air is then rolled out of the system and the moulding allowed to harden.

Hand lay-up techniques do have, however, a number of disadvantages. For example a good finish is normally only possible on one side of the moulding. It is also not suitable for certain shapes, e.g. fishing rods. Thus such systems as *matched-metal moulding*, involving a mould with matching die faces, often in association with pre-impregnated sheet material (sheet moulding compound) (see below) and pultrusion, in which rovings are resin impregnated and then pulled through a shaping die, are used.

Dough moulding compounds (DMC) and sheet moulding compounds (SMC)

Dough moulding compounds were developed primarily in an attempt to combine the excellent mechanical properties of polyester–glass fibre laminates with the mouldability and rapid cure of more conventional thermosetting moulding materials. They are made by blending a liquid resin (usually polyester) with filler, reinforcing fibre (usually glass), pigment, lubricant and hardener. The hardener is selected to be stable at room temperature but very active at conventional compression moulding temperatures. The unreacted mix is like a fibre-filled putty which for convenience is often supplied in rope-like form. As with common polyester laminating resins the DMC compounds cure without evolution of volatiles so that low moulding pressures (c. 7 MPa) may be employed. A number of special problems are associated with DMC. Besides being somewhat inconvenient to handle, even in rope-like form, it may also:

1 exhibit a tendency for thick sections to crack;
2 be difficult to avoid warping and to maintain close tolerances;
3 give an undesirable surface pattern due to fibre being left prominent when the resin shrinks during cure;
4 be difficult to mould large parts with any control over fibre orientation.

The first three of these problems have been minimized by the use of *low profile polyester resins* in which the polyester resin has been blended with a thermoplastics material such as polystyrene or some acrylic polymer.

The last problem has been tackled by the introduction of *sheet moulding compounds*. A typical SMC is made by spreading a mixture of resin, filler pigment and hardener onto polyethylene film. Chopped glass fibre is then spread onto the resin which is then covered by a further polyethylene film coated with the resin mix so that the glass fibre forms a sandwich between the resin mixes on each polyethylene film. The whole assembly is then rolled up for storage and transport. Fillers are selected that tend to weakly gel the resin to give a leathery consistency and easy handling when the sheet materials are put into the mould. The gel is, however, sufficiently weak for the resin to be capable of flow under shear during the processing operation. The system may be adapted to give preferred orientation of the chopped fibres or may use continuous filament or even woven textile material. Sheet moulding compounds have become of particular importance to the automobile industry and also for building components.

MIXING

An important intermediate stage in the manufacture of polymer products is that of mixing. If this stage is carried out inadequately then the final product may be unsatisfactory. In some cases this may simply be poor appearance, but often physical properties and ageing behaviour may also be affected. This section introduces certain concepts of mixing and briefly indicates some of the more important mixing techniques.

The term *simple mixing* is used to describe any operation that increases the randomness of the distribution in space of the particles to be mixed *without reducing their size*. An example would be the mixing of white balls with black balls. The mixing would be considered ideal when the distribution was completely random. A number of statistical tests have been developed in order to assess the degree of mixing.

Where the mixing is accompanied by a reduction in the size of the particles being mixed, for example if the black and white balls were being pulverized during the mixing, the process is described as *dispersive mixing*.

Where liquids and gases are involved there may be a natural diffusion of part of the material being mixed and the term *diffusive mixing* may be used.

Simple mixing as defined above may be important when preparing blends of virgin material with masterbatch granules. Various types of tumble blender may be used for this purpose. It should be noted that where the masterbatch granules have a different density to that of the virgin material some so-called mixers may in fact tend to *reduce* the randomness of the distribution, i.e. act as demixers.

In many mixing processes there is a premixing stage where the components are stirred together under quite modest levels of shear. Some dispersive mixing may occur here as aggregates and flocculates are broken up and the distribution of components begins to become randomized.

In the case of polymers, much mixing is carried out when the polymer is in the melt state. In such circumstances the shear levels are very high and heavy duty machinery is required. This process usually invokes *laminar mixing* (see below).

GELATION OF PLASTICIZED PVC

Since mixing is particularly important with plasticized PVC it is appropriate here to briefly mention the process of *gelation* that occurs when PVC and plasticizers are heated (usually under shearing conditions) converting a damp powdery premix into a tough elastic mass.

The first stage of gelation is to heat the components above the glass transition temperature (T_g) of the PVC (c. 80 °C). This causes the PVC to soften and as the PVC molecules jostle about space is created between the molecules. Liquid plasticizer molecules are then able to diffuse into the PVC particles, which become swollen, facilitating further inward diffusion of plasticizer. If the mass is then pounded together PVC molecules from one particle are able to interpenetrate into adjacent particles and eventually the mass fuses into a homogeneous rubbery mass. The term *fusion* is sometimes used instead of gelation in this context. This is an example of diffusive mixing. If the process is inadequate, complete homogenization is not achieved, and the mix may not show optimum physical properties. When there is gross undergelation the mix may be quite friable. At the same time PVC has limited thermal stability so that too extensive mixing at elevated temperatures will also cause the product to be unsatisfactory. In this context it is particularly important that all the PVC be subjected to similar mixing histories and that no particles or other fragments of the PVC mix become trapped in the mixer and sustain overmixing and degradation.

Note: The use of the word gelation in this context is different to its use in relation to thermosetting plastics. In the latter case it is used to denote the change from a liquid to a solid (particularly a rubbery) state due to chemical cross-linking processes.

LAMINAR MIXING

As mentioned above mixing of laminar melts usually involves laminar mixing. A simple example of this is to consider mixing between black and white material in the space between two concentric cylinders which are rotating with respect to each other (see Figure 5.59).

In Figure 5.59(a) some black material is initially in a straight line across the annular gap. The figure then shows what happens to this line of material after one of the cylinders has turned 1,2,3 and 1000 rotations. Clearly the black material becomes more mixed into the white with each revolution. Numerically if the gap is L then the distance r between adjacent lines of the spiral will be given by

$$r = L/(n + 1) \simeq L/n,$$

where n is the number of turns.

In Figure 5.59(b) the original placement of the black material only extends across part of the gap and mixing will be confined to only part of the white material. In Figure 5.59(c) the original placement of the black material is annular and no mixing takes place.

It will be appreciated that for effective mixing the following criteria will have to be met:

1 The interfacial area between the components must be greatly increased.
2 The elements of the interface must be uniformly distributed.
3 The mixture components must be distributed so that in any (characteristic) volume elements the ratio of the components is the same as that of the whole system.

Only in Figure 5.59(a) do we see these requirements being met. In this system it will be noted that

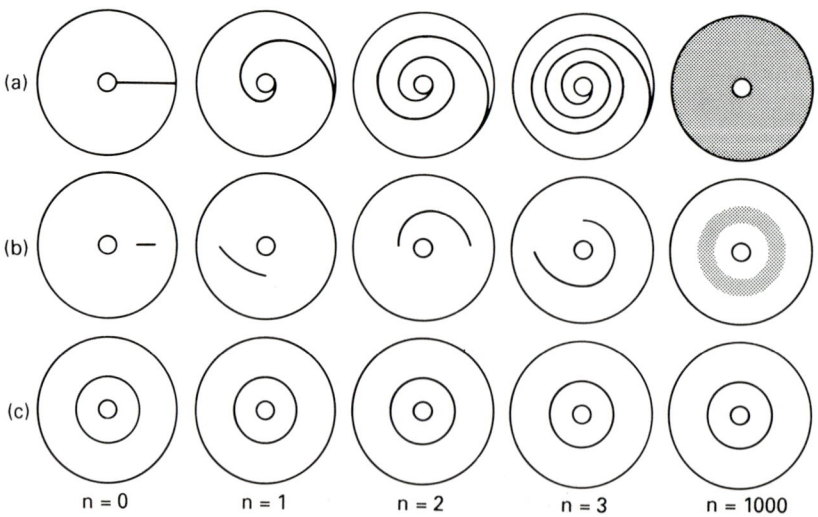

Figure 5.59 *Representation of a laminar, non-diffusive mixing process between concentric cylinders. Initial orientation of components and total shear strain determine the extent of mixing*

1 The initial orientation of the interfaces is at an angle to the flow lines (streamlines) (in this case at right angles).
2 All streamlines cut through the interface.
3 There is shearing between adjacent streamlines.

   Such requirements must be met if any mixing system is to be effective. It may also be noted that in the above simple example there is no axial mixing so that we should add:
4 There should be a mixing component in three perpendicular directions.

MIXING SYSTEMS FOR POLYMER MELTS

Many of the methods used for mixing plastics were adapted from those used for rubber mixing. These included *two-roll mills* which require considerable operator skill and are not efficient for large scale work although useful for laboratories and *internal mixers* (see Figure 5.60) in which the compound is mixed by two heavy duty rotors inside an enclosed chamber. The main disadvantage of internal mixers is that they are not continuous processes. Considerable progress has been made in developing systems that do not have this disadvantage. These are largely based on extruders.

A single screw extruder unmodified is not a good mixer since material in the centre of the channel is subjected to little shear and little laminar mixing between the hopper and the head. If however the flow lines can be broken up and material continually redistributed across the channel then good mixing may take place. Many modified single-screw extruders are now available that function in this way.

Similar deficiencies also occur with twin-screw extruders. Where, however, the screw threads are interrupted by mixing discs or the flow pattern disturbed in some other way, efficient mixing may again occur.

It is, however, important to stress that in a continuous mixing process the feed of components must also be constant. If the composition of the feed varies with time so the composition of the final mix will tend to vary.

CRITERIA OF A GOOD MIX

The criteria of a good mix will vary from application to application. In some cases it may be

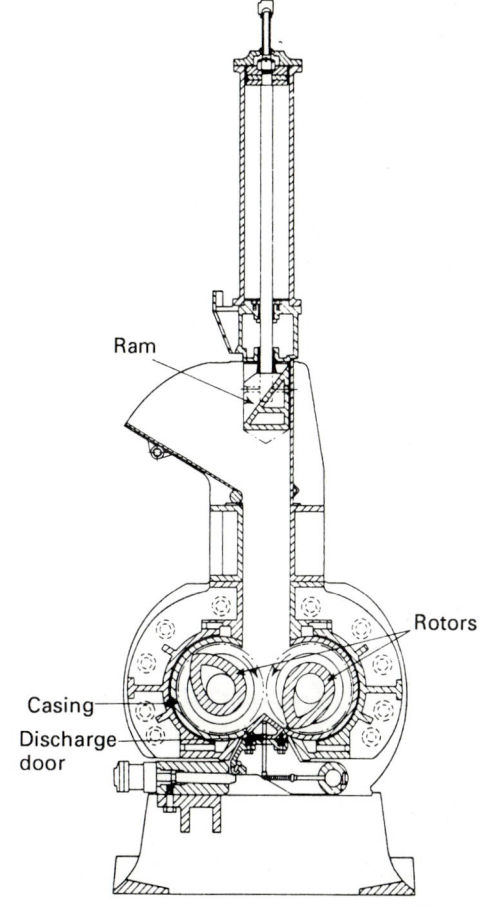

Figure 5.60 *A typical internal mixer (Banbury mixer) widely used for mixing polymers. The mix components are fed down the shute into the mixing chamber. The system is enclosed by a ram which exerts a pressure on the mix as the rotors rotate. At the end of the mixing cycle a discharge door opens and the mix is fed to a two-roll mill or an extruder for further handling*

adequate that the compound appears of even texture to the naked eye, for example where carbon black is being used simply as a pigment. If however carbon black is being used in polyethylene as a UV absorber then experience has indicated that the mix should still appear to be evenly dispersed when viewed at $400 \times$ magnification under a microscope. With plasticized PVC the polymer–plasticizer mixing should approach a molecular scale, best assessed by measuring some physical property such as tensile strength. It may also be necessary, particularly with PVC, to check that the compound has not been subjected to overheating during mixing and that it has an acceptable level of stability.

MASTERBATCHING

In many cases of mixing it is required to use only a small but critical amount of additive, e.g. colourant or antioxidant. When these are in the form of powder it is sometimes difficult to meter this accurately into the mixing equipment. This problem can be reduced by the use of *masterbatches*, polymer containing heavy concentrations of the additive. At the appropriate time, usually in the hopper of the moulding machine or extruder, the masterbatch may be blended with the virgin polymer. This approach also has the advantage that in many operations the manufacturer only needs to store virgin polymer and colour masterbatch rather than large stocks of different coloured compounds.

While blending may be carried out in tumble blenders of various designs, many systems have been developed for metering masterbatch material into virgin polymer at the hopper stage. It would, however, be stressed that these methods usually involve some sort of volumetric metering whereas the formula will be based on weight ratios. The units should therefore be calibrated for each batch. Furthermore any change in the bulk density of the granules (which will be affected by particle shape and size) will also affect the results.

PROCESS POTENTIAL

Comments on the scope and potential of the processes described can only be indicative since continual development of each process progressively increases both their scope and productivity.

Injection moulding is often considered as the ultimate mass production process. Highly complex shapes may be made in one operation which might otherwise have to be made by assembling large numbers of separate parts. Machines are available over an enormous range of sizes with shot capacities ranging from a few grammes to those sufficient to mould small boats. Cycle times with small thin-wall mouldings can be less than a second but with most moulding cycles probably being between 10 s and 1 min. The cycle time depends on the shot size, the material characteristics and the shape of the moulding. With thermoplastics the cooling stage of the cycle depends on section thickness. Excess section thickness can cause cycles to be unacceptably long as well as being excessive in material usage and possibly also causing problems of quality such as warping and sink marks. It is therefore normal to design parts of limited section thickness using ribs and other features to enhance rigidity (but see the note on structural foam materials in the section on product design). In recent years injection moulding of thermosetting plastics has become more common. In this case the shot capacity must not be so great that the mould cavities cannot be filled before the material starts to cure.

Because of their intricacy and need to be robust, injection moulds and injection moulding machines tend to be expensive and the process can normally only be considered where a large number of parts are to be produced.

Extrusion is probably used to process a greater tonnage of plastics than injection moulding, although in part this will be because the process may be used to make intermediate stage materials such as sheet. Extrusion is largely restricted to the manufacture of products of constant cross-section, although post-extrusion shaping can allow some deviations from this. As with injection moulding machines, extruders are available in a wide range of sizes and can produce products ranging in size from the ink reservoirs of ball point pens and fishing lines to huge sewage pipes and tubular film many feet in diameter.

Most extruders have screws of diameters between 30 and 200 mm with the latter capable of giving maximum outputs in excess of 2000 kg/h in the case of some materials such as plasticized PVC. Linear output rates can vary between a few inches per minute in the case of large pipe to well over 1000 ft/min. in the case of wire insulation.

Blow moulding is somewhat more specialized, being used for hollow objects. Whilst used mainly for bottle manufacture, blow moulded products range in weight from a few grammes to over 20 kg, with products ranging from small hollow spheres of 5 mm diameter used to reduce evaporation loss from liquid surfaces, via bottles, jerrycans, dustbins and 200 litre drums to 1 m^3 oil storage containers. For some of the larger applications *rotational moulding* may be considered as a competitive process which generally involves a lower capital outlay.

Compression moulding is largely restricted to thermosetting materials and was at one time the main process of the industry. It has been traditionally associated with fairly small mouldings weighing at most 5 kg and usually much less, although during World War II coffins were made by compression moulding and more recently large mouldings have again been manufactured from sheet moulding compounds. Production rates are limited by the cure rates possible, most moulding cycles being in excess of one minute.

Calendering is restricted to the manufacture of sheet or coating of cloth with plastics and rubbers. It is used mainly for PVC. There are limitations to width of sheet, with maximum roll face widths rarely above 3 m. Most film and sheet produced is in the thickness range of 0.075–0.900 mm, although sheet up to 1.5 mm thick can be produced. The capital cost of installing a modern calender line is generally much higher than for other processes, but because of the possibility of high output rates and good thickness control, the process continues to be used for the manufacture of the bulk of PVC sheeting.

Sheet forming and other thermoforming processes also continue to be widely used. Whilst there are limitations in the range of shapes that may be achieved without having to assemble separate parts, the process is attractive in that quite large mouldings may be made using less expensive equipment and moulds than with injection moulding. It is possible to make small numbers of mouldings using moulds of wood and even plaster. As with other processes, products with a wide size range are possible, from simple blister packs to interior aircraft trim and sides of caravans. Particularly well-known applications include baths, food packs, automobile trim and advertising signs.

Fibre-reinforced composites also range from small injection mouldings to boats more than 40 m in length made by hand lay-up techniques. Some of these techniques enable very large products to be made with low capital costs and whilst they may be slow (with in some cases the moulding being left on the mould for a period of days), the process can be a useful way of handling small runs which would be uneconomic with other processes.

Cellular plastics (which are not dealt with in any detail in this handbook for reasons of space) may be made by a variety of techniques. In volume terms, materials such as the cellular polyurethanes and expanded polystyrene materials comprise a significant proportion of plastics produced, and are of particular importance for insulation and packaging purposes.

6

TROUBLE SHOOTING GUIDE

INJECTION MOULDING

Fault	Cause	Action
Short (incomplete moulding)	Feed problems.	Check feed to barrel. Check feed from nozzle. Check screw travel limit switches.
	Inadequate flow.	Check melt temperature (may be too low). Check injection pressure (may be too low). Check mould temperature (may be too low). If using hot runners, check their temperatures. Check that flow path ratio is not excessive.

Fault	Cause	Action
Flashing	Worn or damaged mould.	Check for damage at parting line. Check for worn mould mechanisms.
	Insufficient locking force.	Increase clamp pressure. Reduce injection pressure. Reduce melt temperature.
Sink marks	Insufficient cavity pressure.	Check packing pressure (may be too low). Check mould temperature (may be too low). Check shot volume (may be too low).
Bubbles/mica marks.	Volatiles/gases. Moisture. Degradation products.	Use dry granules. Avoid overheating melt. Use decompression screw.
	Volatile additives	Avoid overheating. Reformulate if necessary.
	Trapped air (may occur with dry blends).	Use screw with higher C.R.
Poor finish (dull).	Poor tool quality.	Check tool for polish damage, corrosion.
	Insufficient pressure in cavity.	Check injection and packing pressures. Check melt temperature. Check mould temperature. Vary injection speeds.
Cold granules (visible lumps or varying opacity in moulding).	Insufficient melting.	Check cylinder temperatures. Check screw speed and screw back pressure.
Metal contamination	Contaminated feed.	Check raw material. Check hopper magnets or other separating devices.
	Machine wear and tear.	Check cylinder, nozzles etc.
Black marks in moulding.	Contamination.	Check raw material. Check barrel and nozzles.
	Degraded material.	Check for dead spots in system.

Fault	Cause	Action
Black marks in moulding (near blind end).	Air trapped in cavity rapidly compressed and causes burn marks on moulding.	Check air vents clear.
Discolouration near gate.	Excess frictional heating at gate.	Lower injection rates. Lower melt temperatures. Lower mould temperatures. Increase gate section.
Cracked mouldings (often around ejectors).	Too high a cavity pressure.	Check injection and cavity pressures. Check melt temperature. Check mould temperatures.
Mould sticking in cavities.	Too high a cavity pressure. Damaged mould.	As above. Check and repair if necessary.
	Bad mould design	Polish surfaces of cavity. Increase cavity sidewall tapers.
Delamination near gate.	Too high a packing pressure.	Reduce packing pressure at end of cooling stage.
Highly strained mouldings (e.g. cracking in certain liquids).	Excess orientation in moulding.	Raise melt temperature. Raise mould temperature.
Drooling from nozzle.	Failure of nozzle seal.	Check that nozzle seal is not jammed. Check on seal design for polymer used.
Drooling between nozzle and sprue bush.		Check nozzle setting. Check nozzle/sprue alignment. Check nozzle/sprue radii. Check that injection stroke is not premature.
Screw does not return.		Check feed from hopper. Reduce back pressure. Increase screw speed.

BLOW MOULDING

Fault	Cause	Action
Parison too long.	Excess extrusion output.	Reduce screw speed.
	Melt too fluid leading to parison sag.	Reduce barrel temperature. Reduce die head temperature.
Parison too short.	Insufficient extruder output.	Increase screw speed.
	Too low a melt temperature.	Check heaters are functioning.
Parison diameter too large.	Incorrect die selection.	Check.
	Excess die swell.	Raise melt temperature.
Parison diameter too small.	Incorrect die selection.	Check.
	Excess parison sag.	Lower melt temperature.
Excess parison wall thickness.	Incorrect setting of parison thickness controller.	Check.
	Excess die swell.	Raise melt temperature.
Parison wall too thin.	Incorrect setting of parison thickness controller.	Check.
	Excess parison sag.	Lower melt temperature.
Parison curls outward as it leaves the die.	Die temperature too low.	Raise die temperature.
Parison curls inwards as it leaves the die.	Die temperature too high.	Lower die temperature.
Parison veers to one side.	Die and mandrel not concentric. Incorrect setting of parison wall thickness control.	Adjust.
Sharkskin (rough on inside).	Die head temperature too low.	Raise die head temperature.
	Melt temperature too low.	Raise melt temperature.

Fault	Cause	Action
Longitudinal marks on parison surface.	Dirty die.	Clean.
	Damaged die.	Repair.
Weld lines on parison.	Melt not re-welding after passing through spider.	Build up back pressure between spider and die lips by lowering die/die head temperatures. Increase land length of die. Increase melt temperature.
Burn marks (PVC).	Decomposition.	Check melt temperature. Check no dead spots. Check tip of screw not too hot.
Dark particles.	Contamination Decomposition	Check. Check dead spots.
Poor neck/flash separation.	Blow pin badly set. Damaged cutting sleeve.	Check and adjust. Replace.
Displaced punt weld.	Bent parison. Blunt parison knife.	Check parison faults. Replace.
Deformed moulding.	Moulding ejected too hot.	Increase cooling time. Check cooling water supply to mould.
Tail flash on bottle.	Short parison.	See parison faults.
Moulding does not fully inflate.	Insufficient air.	Raise air pressure. Increase blowing time.
Moulding sticks to mould.	Mould too hot. Moulding too hot.	Check mould temperatures. Increase cooling time. Reduce melt temperatures.

COMPRESSION MOULDING

The following faults apply to PF and aminoplastic (UF or MF) moulding materials and not necessarily to other thermosetting materials. They would not in general apply to compression moulding of thermoplastic materials.

Fault	Cause	Action
Short moulding; thin flash.	Insufficient powder.	Increase charge.
Short moulding; thick flash.	Premature cure before flow complete due to: Too high mould temperature.	Check and reduce if necessary.
	Insufficient pressure.	Increase.
	Inadequate flow.	Change to freer flowing powder.
	Incorrect mould setting.	Correct.
Thick flash on otherwise good moulding.	Excess of powder.	Reduce charge.
Blisters	Undercured moulding.	Increase cure time.
	Mould temperature too high.	Reduce.
	Trapped gas.	Increase preheat temperature. Breathe mould.
Ripple on deep draw moulding or orange peel surface on shallow moulding.	Wrong preheat temperature.	Adjust.
	Uneven flow in mould.	Close mould more slowly. Use freer flowing powder.
	Wrong mould temperature.	Adjust.
Poor finish on flat surface.	Incomplete consolidation.	Use finer powder. Use stiffer grade of powder.
'Burning' (aminoplastics) (white blisters, bleaching).	Overcure.	Reduce cure time. Reduce moulding temperature.
Whitish bloom or small white specks (aminoplastics).	Precure	Reduce preheat time and/or temperature.
	Powder charge in cavity for too long before pressing.	Apply pressure sooner after charging mould.

EXTRUSION

A wide range of plastics is commonly extruded to make such different products as rod, tube, film, sheet, solid profiles and wire coverings. For each material and each product type there are specific causes of faults and remedial procedures. This section is therefore intended as a general approach to extrusion trouble shooting.

CAUSES OF EXTRUSION FAULTS

1 Contamination or use of unsuitable raw material.
2 Incomplete melting of polymer (often also known as incomplete plasticization of the melt).
3 Overheating of melt.
4 Rheological phenomena such as melt fracture and sharkskin.
5 Use of unsatisfactory equipment.
6 Incorrect die settings.
7 Unsatisfactory cooling and haul-off arrangements.

MAIN TYPES OF FAULTS (THE NUMBERS REFER TO THE POSSIBLE ORIGINS OF THE FAULTS AS ABOVE)

- Extrudate not homogeneous in colour (1, 2, 3)
- Extrudate rough (2, 3, 4)
- Extrudate tends to distort on emerging from die (1–6)
- Dimensions of extrudate incorrect (2, 5, 6, 7)
- Longitudinal marking on extrudate (5, 6, 7)
- Poor mechanical properties (1, 2, 3, 7)
- Poor optical properties (e.g. lack of clarity in thin film) (1, 2, 3, 4, 7).

TYPICAL FAULTS AND THEIR REMEDIAL TREATMENT

1 A solid profile is being extruded in which the cross-section varies in thickness. The extrudate tends to coil as it emerges from the die (or possibly may tend to ripple at one point on the surface. This indicates that flow is faster through one part of the die cross-section. If the die has been working satisfactory a blockage is indicated towards the rear of the die which may be due to contamination, overheating by degradation or possibly incomplete melting. If the job is a new one it indicates faulty die design or die setting. Depending on the design in question, solutions should be sought in either reducing the die parallel to where the flow is slowest, or building in some constriction to flow (e.g. by means of a dam or a pin) where the flow is fastest.

2 An extruded tube has a smooth outside surface but the inner surface shows transverse rippling. This is a result of the fact that in an annular die the maximum melt velocity is nearer to the inner than the outer wall. Where the problem is serious it may often be solved by building in a restrictor bulge onto the central pin to slow up flow in this area.

3 Large bore tube shows transverse ridges on the outer surface. This is probably due to 'sharkskin', a tearing of the melt on the outer surface of the extrudate as it is pulled away from the die. This effect occurs above a critical linear extrusion rate and thus disappears if the extrusion speed is sufficiently reduced. An increase (or sometimes even a decrease) in the melt temperature may increase the critical rate whilst die tip cooling may also be advantageous.

4 Wire covering is uneven with some helical distortion. The most likely cause is melt fracture, a distortion which occurs above a critical shear rate (as opposed to a critical linear output rate). The critical shear rate may be increased by raising the temperature whilst tapering of both the die entry and the die 'parallel' can also be beneficial. The problem tends to be greater with higher molecular weight polymers and with polymers of low solubility parameter.

5 Lay-flat film has unbalanced mechanical properties and a lower impact strength than might be expected. This may be overcome by ensuring that the molecules are stretched equally in the machine and transverse directions just before the melt freezes. To some extent a bubble of shape B is more likely to achieve this than one of shape A. Bubble shapes may in turn be controlled by varying

extrusion/haul off rates and varying the extent of cooling.

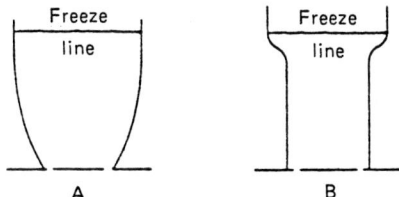

Molten polymer emerging from an annular die
A most of the lateral expansion complete well before freezing
B Lateral expansion occurs just before freezing.

6 Polyethylene lay-flat film is not sufficiently clear. Providing there is no contamination, the cause will be either due to surface roughness due to melt unevenness not having time to disappear before the film freezes, or to crystallization because the time before the film freezes is too long. A compromise on cooling time is thus necessary, but other factors such as cooling air supply, haul-off speed, blow ratio and melt temperature can also be adjusted to minimize the opacity.

VACUUM FORMING

Fault	Cause	Action
Torn/split mouldings.	Corners too sharp.	Increase corner radius. Increase section taper.
	Moulding too cold on ejection.	Increase mould temperature. Reduce cooling time.
Chill marks (sudden change in thickness).	Sudden chilling of sheet in contact with cold surface (mould or plug).	Warm mould or if due to cold plug in plug assist technique may be possible to warm plug, grease the plug or blow warm air out of plug to form a lubricating air layer.
Webbing between high points, usually in mouldings and particularly with multi-cavity moulds.	High points too close.	Consider possibility of female moulding. Use of skeleton tooling or plug assist. Increase taper and radii. If possible redesign to increase distance between high points and reduce height. Evacuate air more slowly. Increase material thickness.

(If the sheet is anisotropic, as may happen if it has been calendered, the problem may sometimes be reduced (or indeed aggravated) by rotating the sheet through 90°.

Orange peel.	Tiny trapped air bubbles on surface of shallow moulds.	Increase venting if acceptable. Sand blast surface to give matt finish (it is sometimes possible to use a layer of fine fabric instead).
Whitening	A stress phenomenon.	Raise shaping temperature.

Fault	Cause	Action
Warping of mouldings.	Strains arising through temperature differences on removal of moulding.	Increase cooling period. Preheat frame and associated metalwork.
Blow-holes, pinholes and bubbles.	Overheating causing degradation. Overheating causes sheet to melt and collapse. Blow holes due to too large vent holes.	Reduce heating time. Reduce vent hole diameters.

CALENDERING

The bulk of calendering operations carried out on plastics involves PVC and this is the material considered here. The process of *gelation* refers to the effect of heating polymer and plasticizer together so that plasticizer molecules diffuse into the polymer particles to give a polymer-plasticizer mixture on a scale of dispersion that approaches a molecular level. Whilst unplasticized PVC (UPVC) is hard, rigid plasticized PVC (PPVC) may be flexible and even rubbery if sufficient plasticizer is present. The process of gelation is usually carried out at the same time as the mixing of the PVC compound, which will also contain many other ingredients. In the following table the heating of the compound to soften it and make it capable of flow will be referred to as *fusion*. Some writers also ambiguously refer to this process as gelation or as plasticization. The reference to gel-free PVC is to polymer that does not contain cross-linked, branched or very high molecular weight material.

FAULTS IN APPEARANCE

Fault	Cause	Action
Fish eyes (nibs).	Undispersed PVC particles.	Increase mixing/gelation time. Use gel-free PVC.
Fleck marks.	Incorrect lubrication.	Reduce external lubricant. Increase mixing/gelation time. Adjust calendering temperatures.
Plug marks.	Poor release of sheet from hot rolls.	Increase lubrication. Reduce roll temperatures. Avoid overheating during mixing.

Fault	Cause	Action
Heat lines along machine direction.	As for pluck marks.	As above.
Surface roughness.	Incomplete fusion.	Raise roll temperatures.
Bank marks (areas of slight roughness like a watermark on paper).	Uneven stock temperature.	Change size of stock bank. Check variation in feedstock temperature.
Overall discoloration.	Polymer degradation.	Increase stabilization. Reduce overheating.
Dark specks.	Contamination. Degraded material.	Check mixing stage. Check mixing stage.
Crows feet (pine trees) (V-shaped marks on sheet).	Inadequate preheating. Too large a feed dolly. Poor dispersion of particulate additives.	Increase preheat temperature. Reduce. Check mixing process.
Pinholes, blisters, windows.	Air trapping. Incomplete gelation. Undispersed particles that drop out.	Reduce bank size. Raise bank temperature. See 'fish eyes'. Check mixing stage.

FAULTS IN MECHANICAL PROPERTIES

Fault	Cause	Action
Poor physical properties.	Insufficient fusion, gelation or additive dispersion. Degradation of polymer.	Increase mixing times, gelation temperatures, calendering temperatures. Opposite to above.
Anisotropy (e.g. higher strength in machine direction).	Overstretching sheet on release from rolls.	Unlikely on modern equipment.

DIMENSIONAL FAULTS

These include such faults as incorrect thickness, transverse thickness variation and longitudinal thickness variation. With modern equipment these faults arise from incorrect setting.

CHAPTER

7

PRODUCT DESIGN PRINCIPLES AND PROBLEMS

CHARACTERISTICS OF PLASTICS MATERIALS

The present author has written elsewhere 'Successful product design requires knowledge, intelligence and flair. The knowledge requirement may in turn be subdivided into:

1 A knowledge of the requirements of the product.
2 A knowledge of the behaviour of plastics materials.
3 A knowledge of plastics processes.
4 A knowledge of all relevant economic and psychological factors.

Intelligence is required to relate this knowledge and flair to bring the design to a successful reality.'

There have been in the past, and continue to be, many failures in the use of plastics which have influenced the public's view of these materials. In some cases the fault is simply due to the wrong choice of material but more commonly it results from the failure of the designer to appreciate some basic properties of plastics materials. In particular mention should be made of the following characteristics of thermoplastics materials. The list may appear formidable, but proper attention to product requirements, material selection and an understanding of the shaping process can enable very satisfactory products to be obtained.

1 Since thermoplastics consist of long chain polymer molecules these molecules become oriented during shaping operations. This can result in the product being anisotropic, i.e. it has different properties in different directions. It can also lead to differential shrinkage on cooling which in turn may lead to warping. This effect may be aggravated by the use of anisotropic fillers such as glass and carbon fibres. Because an oriented molecule is not in a naturally stable state (it will try to coil up in a random fashion) internal stresses may be set up in the product. This may mean that the additional amount of external stress that has to be imposed before the product breaks will be less than if there is no internal stress. (In some cases orientation will increase the strength, particularly where the strength is measured along the direction of orientation.)
2 Shrinkage on cooling may be quite substantial, particularly with crystalline polymers. (In this case shrinkage will be reduced by the presence of glass fibre.) If the products have thick sections and because it is the surface which is usually the first to cool, shrinkage may be 'outward' towards the surface and this can result in voids or, alternatively or additionally, sinking at the surface.
3 Polymer melts are surprisingly quite compressible, up to 8% compression not being uncommon during injection moulding. Overpacking of polymer molecules into a mould may lead to severe problems of moulding ejection.
4 At sharp edges and at sudden changes of cross-section, high concentrations of stress can occur, providing a weak point in the product making it liable to premature failure, particularly in some aggressive environments. Amongst aggressive environments are liquids that do not dissolve the polymer but rather tend to soften the surface. Generally speaking these materials will have a solubility parameter (q.v.) just outside the solubility range.
5 In the case of injection moulding, the melt is cooling as the mould fills and if the flow path ratio (flow length/flow cross-section) is too

great a mould cavity may not fill. Whilst filling may be possible if the injection pressure is increased, this may cause overpacking at the gate end of the moulding causing release problems.
6 The properties of plastics are generally more dependent on temperature than are traditional materials such as metals and ceramics.
7 Many thermoplastics will tend to creep substantially under low loads making them unsuitable for load bearing applications.
8 Many plastics change in their properties with time due to ageing. In many cases ageing may be accelerated by heat, oxygen and ultraviolet light.

The product designer may have to consider a range of properties in order to assess the suitability of a particular plastics material to his purpose. The three properties discussed below are important for a very wide range of applications, whatever the shaping process.

POLYMER RIGIDITY AND EFFECT OF TEMPERATURE ON RIGIDITY

In the case of a *polymer*, as opposed to a filled plastics material, the rigidity of the material is determined by the ease with which the polymer molecules are deformed under load. At $-273\,°C$ all load is taken up by bond bending and stretching and for a polymer with no secondary transitions this state of affairs will hold up to the region of the glass transition temperature (T_g). There are, however, several polymers in which small parts of the molecule start to move below the T_g and these may respond to the application of stress, although changes are generally small. At the T_g, segments of the molecule become quite flexible and the polymer mass will deform quite easily under stress. For example, in a typical amorphous thermoplastics material the modulus will drop from a value of the order of 3500 MPa below the T_g, to less than 0.7 MPa at a few degress above it. In the case of an unfilled amorphous material of only moderate molecular weight further raising of temperatures above the T_g will rapidly lead to melting and the modulus will approach zero. (See Figure 8.7, page 142.)

In the case of crystalline polymers the modulus may drop at the T_g but the drop will depend on the level of crystallinity, with a smaller reduction in the case of highly crystalline materials. Between T_g and T_m the modulus will drop slowly but then drop rapidly around the T_m. Fibre-reinforced thermoplastics tend to retain a high modulus until close to the T_m. (See also Figure 8.7.).

At this stage, it is worthwhile distinguishing between the rigidity of the plastics material and the stiffness of a plastics component. As will be seen later in the chapter, there are a number of useful ways of increasing stiffness independently of the material used.

TOUGHNESS

Plastics components are rarely used under loads approaching their ultimate strength, be it in tension, compression, flexure or shear. Far more important in practice is fracture due to impact.

Some standard impact tests are described in Chapter 8 but in themselves such tests cannot be expected to tell whether a particular product will perform well in service without fracturing or permanently distorting on impact. In practice the designer should consider the nature of likely forms of impact in service and try to design the part and its method of manufacture in order to overcome likely problems. For example, it is shown in Chapter 8 that where a component has a flat surface in which the polymer is highly oriented, the component will have a lower impact resistance than when the molecular orientation is low. On the other hand, if impact is to be expected on a rod-like component or section of a component, then the greatest impact resistance will occur in highly oriented materials. Basically, if a crack has to cross many polymer molecules the energy to break will be much higher than if the crack can grow *between* polymer molecules. Fibrous fillers, providing they are well bonded to the polymer can have a similar effect.

For any critical application it is common to devise impact tests for the particular product.

Examples are a falling pendulum test for whole truck cabs which will simulate a head-on collision, drop impact tests for battery cases and beer bottle crates and high speed projectile impact tests on, for example, protective eyewear and lawn mower hoods.

TOUGHNESS AND THE FRACTURE MECHANICS APPROACH

Traditional impact tests on plastics materials involve rapid loading of standard samples under standard conditions and thus give some measure of the relative strengths of different materials. They cannot however indicate the potential behaviour in service of a manufactured part with definite geometries in a specific stress environment. Considerable progress has been made in recent years to remedy this deficiency by the development of the fracture mechanics approach for use with polymeric materials. This section is intended as an introduction only to this highly complex subject.

Modes of failure
Standard fracture mechanics tests recognize three modes of crack growth. These are:

- Mode I an opening or tensile mode.
- Mode II a sliding shear mode.
- Mode III a tearing shear mode.

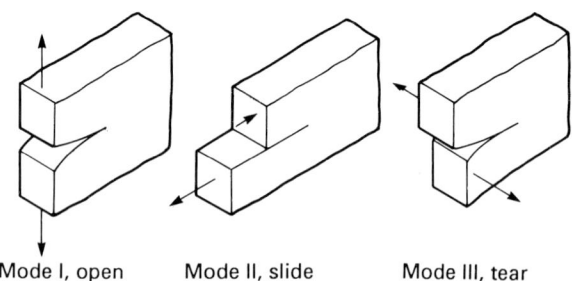

Mode I, open   Mode II, slide   Mode III, tear

Figure 7.1 *Modes of crack growth*

Only Mode I will be discussed in this section, during which the suffix I will be attached to some symbols to denote that they refer to Mode I operations. The discussion is also restricted to the so-called linear elastic materials in which stress is proportional to strain.

The stress intensity factor (for Mode I)

This concept is introduced by first considering an infinitely large thin plate which has an internal crack of length $2a$ and a straight front through the thickness. The plate carries a unidirectional stress σ perpendicular to the crack and which is uniform remote from the crack. This stress tends to open up the crack (see Figure 7.2(a)).

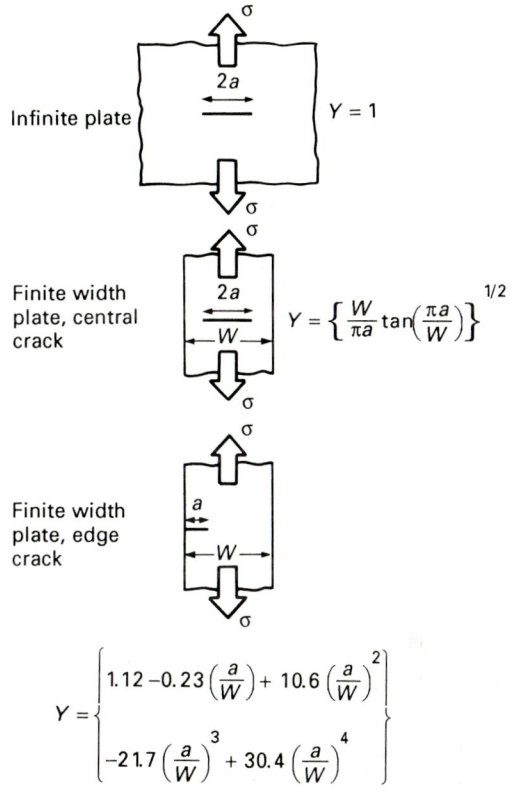

$$Y = 1$$

$$Y = \left\{ \frac{W}{\pi a} \tan\left(\frac{\pi a}{W}\right) \right\}^{1/2}$$

$$Y = \left\{ \begin{array}{l} \left[1.12 - 0.23\left(\frac{a}{W}\right) + 10.6\left(\frac{a}{W}\right)^2 \right. \\ \left. -21.7\left(\frac{a}{W}\right)^3 + 30.4\left(\frac{a}{W}\right)^4 \right] \end{array} \right\}$$

Figure 7.3 *Finite plate correction factors for different geometries*

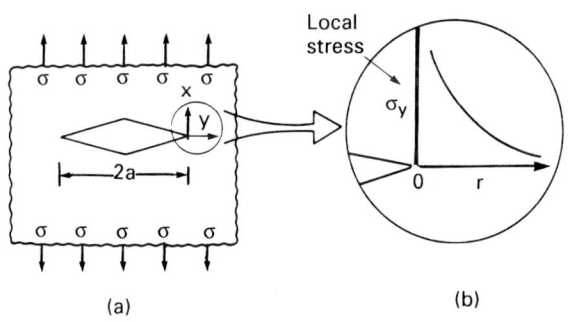

Figure 7.2 *Stresses near crack tip in infinite thin plate*

Near to the crack there will be non-uniform stress with local stresses σ_y *near to but not at* the crack tip which will depend on the distance ahead of the crack tip by the relationship

$$\sigma_y = K_I / \sqrt{(2\pi r)}. \qquad (7.1)$$

This equation may be rewritten

$$K_I = \sigma_y \sqrt{(2\pi r)}. \qquad (7.1a)$$

The constant K_I is known as the *stress intensity factor* (for Mode I) and is defined by the equation

$$K_I = \sigma \sqrt{(\pi a)} \qquad (7.2)$$

K_I is independent of the material (providing the material is linear elastic and isotropic). Consideration of the above shows that the local stress in the x direction ahead of the crack will depend on the applied stress σ and the crack length $2a$ (Figure 7.2(b)).

For *real* situations as opposed to a theoretical thin plate of infinite area a correction factor Y has to be incorporated into the equation so that it becomes

$$K_I = Y\sigma \sqrt{(\pi a)} \qquad (7.3)$$

Values for Y for various geometries are given in a number of standard engineering handbooks with those for two comparatively simple geometries being shown in Figure 7.3

Fracture toughness

Brittle fracture occurs almost instantaneously when K_I reaches a critical value K_{IC}. This corresponds to a *combination* of applied stress and crack length just before the instant of fracture.

The value of K_{IC} is a *material property*, i.e. it depends on polymer type, test conditions and on the processing history of the material. Some typical figures for K_{IC} in air at 20 °C are given in Table 7.1.

Table 7.1

Material	K_{IC} (MN/m$^{3/2}$)
Epoxy resin	0.6
Polystyrene (GP)	1.0
Acrylic sheet (cast)	1.6
Polycarbonate	2.2
UPVC pipe compound	2.3
HDPE pipe compound	3.0

The fracture toughness approach is attractive in that providing reasonable assumptions may be made about permissible crack lengths and possible flaw sizes the K_{IC} can have a much more general application than the use of brittle strength alone.

Many polymers exhibit a measure of plastic yield near the crack tip. Such yielding may make the polymer much more resistant to crack growth and by resulting in a non-linear stress–strain curve may render the linear elastic mechanics fracture mechanics (LEMF) approach invalid. Whether or not this does so depends on the size of the plastic zone ahead of the tip.

It is sometimes assumed that if the length of the plastic zone ahead of the tip is b and that if $b < a/2$ then the LEMF approach may be used, based on an effective crack length $a_{eff} = a + b$. Where $b > a/2$ then the LEMF approach cannot be used.

For materials that yield, the local stress σ_y cannot exceed the yield stress σ_Y. This means that up to a distance b ahead of the tip the maximum local stress will be σ_Y. Substituting this value into equation (7.1a) we obtain

$$b = K_I^2/(2\pi\sigma_Y^2) = a\sigma^2/(2\sigma_Y^2). \qquad (7.4)$$

This yield zone, or plastic zone is often observed as a white voided area.

Crack growth rates

Under a constant load it has been found for many polymers that the crack length a increases with time t at a rate da/dt related to the stress intensity factor by the expression

$$da/dt = C_1 K_I n \qquad (7.5)$$

where C_1 and n are constants. Since $K_I = Y\sigma \sqrt{(\pi a)}$ it follows that bigger cracks grow faster than small ones.

The time taken for a crack to grow from a_1 to a_2 is obtained by integrating this equation; the solution being

$$t = 2 (a_1^{1-n/2} - a_2^{1-n/2})/[(n-2)C_1 (Y\sigma \sqrt{\pi})^n]$$

In practice, for slow growth n is usually large and in the range 7 to 25 from which it may be shown that the time taken is dominated by the initial crack size.

Use of the LEMF approach

The LEMF approach is particularly useful for predicting the life of parts such as pipes and beams which are found to possess small flaws and cracks.

The following example is taken from *Engineering with Polymers* by P. C. Powell, Chapman and Hall, London (1983). This book also contains a useful list of text books on fracture mechanics.

Problem: At what proportion of the yield pressure would you expect short-term failure in a thin-walled plastic pipe if the depth of an external surface flaw is 0.5 mm? For the plastics material used the yield stress is 60 MPa and the fracture toughness is 2 MN/m$^{3/2}$.

Solution:

Substituting for K_I and σ_y into equation (7.4) we have

$$b = (2/60)^2/2\pi = 1.77 \times 10^{-4}.$$

Combining equations (7.2) and (7.4) we obtain

$$\sigma^2/\sigma_Y^2 = (K_I^2/\pi a)/(K_I^2/2\pi b).$$

From which

$$(\sigma/\sigma_Y)^2 = 2b/a = 2 \times 1.77 \times 10^{-4}/(5 \times 10^{-4}).$$

And in turn

$$p/pY = \sigma/\sigma_Y = 0.84$$

CREEP

When plastics materials are subjected to long term loading they tend to stretch or *creep* in the direction of the stress. This may be due to uncoiling of the polymer molecules (in which case recovery will eventually occur after release of stress) or slippage of chain or crystal structures past each other (when the deformation will be permanent). In the very long term, chemical changes in the polymer leading to degradation may occur, which will cause the material to soften and extend under load.

Whilst there have been many theoretical studies of creep in practice, the designer will find it best to use an *ad hoc* system known as the *pseudo-elastic design approach* using creep curves supplied by the plastics material manufacturer. These will give the extent of creep under a variety of loads, preferably over a range of temperatures, and in the case of hygroscopic plastics, over a range of humidities (Figure 7.4).

In some instances it may be more convenient to use *isochronous stress–strain curves* (Figure 7.4(b)) or *isometric strain–time curves* (Figure 7.4(c)), both of which may be derived from the original family of creep curves.

In outline the *pseudo-elastic design approach* consists of the following steps:

1 Ascertain the function and service conditions of the component part including the expected lifetime and maximum service temperature.
2 Before commencing calculations assume the *worst case*, e.g. assume that the component operates continuously at the maximum service temperature and under the maximum load encountered during its service life.
3 Select the appropriate formula from classical elastic analysis.
4 Select the appropriate figure for stress, modulus, etc, from the creep curve or a derivative of it. Insert this into the formula.

The following example is taken from *Plastics Materials* by the author of this handbook.

A blow moulded container, cylindrical in shape but with one spherical end is prepared from the polysulphone material whose creep curves at 20 °C are shown in Figure 7.5. The cylindrical part of the container has an outside

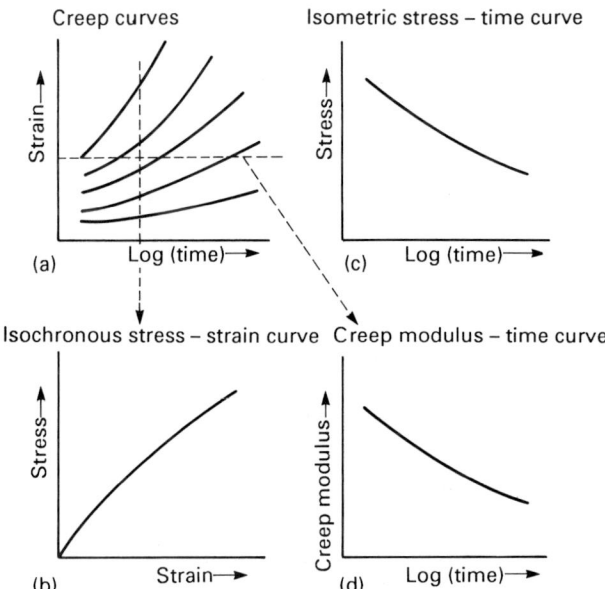

Figure 7.4 *Presentation of creep data; sections through the creep curves at constant time and constant strain give curves of isochronous stress–strain, isometric stress–log (time) and creep modulus–log (time). From ICI Technical Service Notes PES 101, reproduced by permission of ICI*

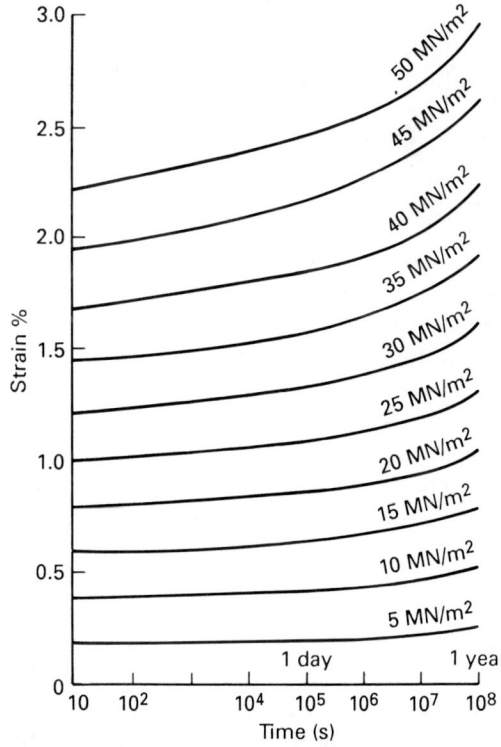

Figure 7.5 *Curves for creep in tension of a commercial polysulphone (Polyethersulphone 300P-ICI) at 20°C. From ICI Technical Service Note PES 101, reproduced by permission of ICI*

diameter at 200 mm and is required to withstand a constant internal pressure of 7 MPa at 20 °C. It is estimated that the required service lifetime of the part will be one year and the maximum allowable strain is 2%. What will be the minimum wall thickness for satisfactory operation?

The appropriate formula from classical elastic analysis is

$$t = pd/2\sigma$$

where t is the thickness, p is the internal pressure, d is the outside diameter and σ is the hoop stress.

Figure 7.5 shows that the stress that will lead to a creep strain of 2% after one year is about 39 MPa. Substituting this into the equation as the hoop stress will give

$$t = 7 \times 200/2 \times 38 = 18.4 \text{ mm}$$

PERSISTENT DESIGN PROBLEMS

There are a number of problems that recur frequently when making components by injection moulding.

SHRINKAGE, FLOW MARKS AND VOIDS

Plastics materials shrink during moulding. Whereas with amorphous plastics moulding, shrinkage is of the order of 0.005 cm/cm the figure is higher with crystalline plastics and this varies with processing conditions, particularly melt and mould temperatures. Some typical shrinkage figures are given on page 182. of the data section.

Where the specification for dimensional tolerance is rigorous, it will be necessary to take into account shrinkage and swelling that might occur due to absorption of water (or other liquids that might be encountered in service). For example if it is expected that the moulding will be used at 65% relative humidity, then the amount of swelling to be expected under such conditions of humidity must be taken into account and will offset the moulding shrinkage figure.

If a thick component is being moulded the centre of the moulding may remain molten long after the surface has frozen. If the moulding is taken out of the mould whilst the centre is still soft and not constrained in a jig, distortion of the moulding may occur due to differential shrinkage.

Another effect is that because the skin is frozen further shrinkage can only occur by pulling in the surface towards the centre, causing *sink marks*, or by microscopic *voids* growing into large visible bubbles which *may* be unsatisfactory. Thus for solid mouldings it is common practice to keep the cross-section of the moulding below critical values above which such problems tend to occur.

Table 7.2 gives some recommended values for minimum and maximum section thicknesses for a selection of thermoplastics. It will be noticed that greater section thicknesses tend to be tolerable with the amorphous polymers.

Table 7.1

Thermoplastic material	Thickness (mm)	
	minimum	maximum
Acetal	0.40	3.2
ABS	0.75	3.2
Acrylic	0.65	6.35
Cellulosics	0.65	4.75
Nylon	0.40	3.2
Polycarbonate	1.00	9.5
Polyethylene (L.D)	0.50	6.35
Polyethylene (H.D)	0.90	6.35
EVA	0.51	3.2
Polypropylene	0.65	7.5
Polysulphone	1.00	9.5
Modified PPO (Noryl)	0.75	9.5
Polystyrene	0.75	6.35
PVC (unplasticized)	0.65	9.5

Thicker section mouldings are possible where a blowing agent is injected with the polymer melt and decomposes to give off a gas, which not only expands the moulding but generates an internal pressure to prevent sinking. In the simpler processes, which yield what are often referred to as *structural foams*, a poor finish will ensue. Good finishes are however possible with *co-injection* or *sandwich moulding processes*, where a foamed core is surrounded by a skin of solid material melted in a separate cylinder.

An alternative process is one in which a gas is injected into the nozzle of the injection moulding machine. The bubbles of gas follow the lines of least resistance and produce cavities in the final moulding, but the gas in these cavities generates an internal pressure usually at points most liable to show collapse. One such process is the Cin-press process.

THE NEED FOR INCREASED PRODUCT RIGIDITY

In the previous section it was shown that unless steps are taken to generate an internal pressure in the moulding by the use of gases, there are strict limits to the thickness that may be obtained with a moulding which is free from sink marks. Such limits in thickness will in turn reduce the rigidity of components consisting of rods, bars or flat plate-like surfaces.

The rigidity of a part that consists essentially of a flat plate may be substantially increased by moulding ribs onto the back of the plate (Figure 7.6). Depending on the likely mode of flexure of the part in service the ribs may be all in one direction or crossed (not necessarily at right angles). An alternative method of stiffening such a part would be to change the shape from a flat surface to a shallow dome if this were acceptable in the design.

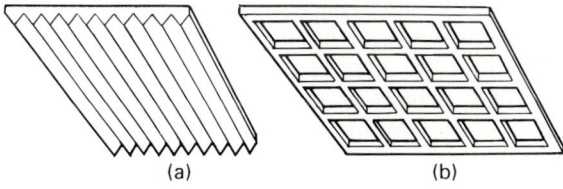

(a)   (b)

Figure 7.6 Increasing stiffness by use of ribs. (a) Stiffness increases in one direction. (b) Stiffness increases in two directions

If neither of these suggestions is acceptable then the use of structural foam moulding should be considered. One particular advantage that may be gained here is that with a structural foam the moulding will be thicker but will have a greater rigidity than a moulding of the same weight produced from a solid plastics material. An example of this is given in the section on specimen calculations (Appendix 1).

ORIENTATION AND DISTORTION IN MOULDINGS

When polymer melts are injected into a mould cavity the individual molecules become distorted, tending to orient with the direction of flow. In a normal moulding process the melt will freeze, at least near the walls, before the molecules have had a chance to coil up again into a random shape. This will make the moulding *anisotropic*, i.e. it will have different properties in different directions. The extent of this difference is often not appreciated, particularly in respect of properties such as tensile strength and flexural strength. This can be demonstrated using the plaque mould shown in Figure 7.7. In a typical experiment, mouldings were made from general purpose polystyrene; in some cases using a centre gate and in others a corner gate.

The mouldings were then cut up into strips, as shown in Figures 7.7(b)–(e) and were broken in flexure. In the case of the centre gated mouldings the samples cut along the line of flow were about 2.5 times stronger in flexure than those cut across the flow lines. In the corner gated moulds the flexural values were similar. *This should not however be taken to assume that there was no orientation. If it had been possible to cut and test samples at 45° to the edges of the plaque differences similar to those obtained with the centre gated plaque should be expected.*

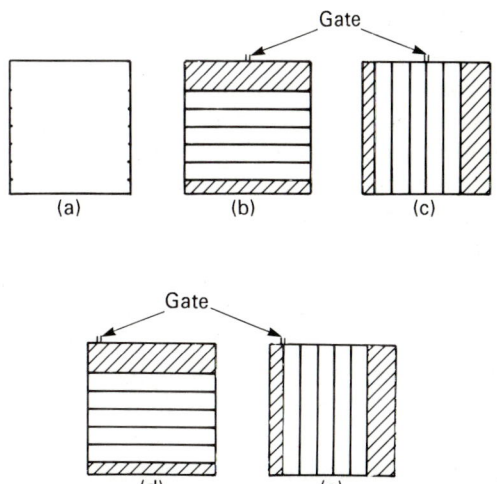

Figure 7.7 Plaque mould fitted with centre gate (b), (c) and corner gate (d), (e). Samples out from (b) have much lower flexure strength than those cut as in (c). In the case of the corner gate moulding the average flexural strengths were similar

More important is the effect on impact properties. This has already been briefly discussed earlier in this chapter and also in Chapter 8). Suffice it to point out here that frozen-in orientation tends to reduce impact strength on flat surfaces but increases it with end gated rod-like shapes.

Orientation is a major cause of distortion in mouldings. If a moulding is withdrawn from the mould before the core has hardened, as is often the case, further cooling may give the molecules in the molten part time to coil up further, exerting stresses in the moulding and possible distortion. Distortion may also occur if the moulding is deformed on removal from the mould cavity, during which time some of the molten core hardens whilst the moulding is deformed, thus freezing-in some distortion.

A further cause of distortion may occur, particularly with crystalline polymers where parts of the moulding cool at different rates (for example where there are different section thicknesses). In such circumstances there may be different total shrinkage which will again tend to distort the moulding.

In general distortion of the mouldings may be minimized by the following actions:

1 Give time for the whole moulding to set before extraction from the mould.
2 Keep cross-sections as small as possible.
3 Keep cross-sections as constant as possible.
4 In the case of crystalline polymers use grades containing crystal nucleating agents which give rapid and more even crystallization.
5 Use an ejection system which minimizes the distortion on ejection from the mould.
6 Use jigs to hold the moulding in the correct shape for a period after manufacture.
7 Take care in stacking mouldings. This may trap heat and allow distortion to occur after moulding.

STRESS CONCENTRATORS

It is well known that if a rod, of wood, glass, metal or plastics material is notched then the strength of the rod, particularly under impact, is reduced. This is explained in terms of high levels of stress being concentrated around the tip of the notch. The sharper the notch angle

and the smaller the notch radius the weaker is the part. Where there are sudden changes in shape in a moulding a situation similar to that of a notch will arise. Some examples of stress points are shown in Figure 7.8.

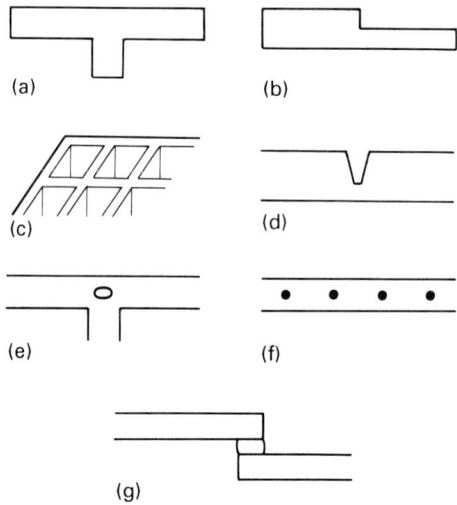

Figure 7.8 *Stress concentrations. (a), (b), (c) Sharp internal angles. (d) A notch (an extreme case of a sharp internal angle). (Materials vary in their notch sensitivity.) A scratch may have the effect of a small notch. (e) Void. (f) Undispersed aggregates of filler, other solid additive or contamination. (g) Cemented or welded joint. The strength of the joint will depend on the quality of the joint*

FLOW AND PACKING

Different materials vary in their ease of flow. This ease of flow is affected by shear rates and temperatures used in the process. In injection moulding, the process is further complicated by the fact that the material starts to cool as soon as it enters the mould cavity, and the material will only be able to flow a finite distance before freezing occurs. One somewhat crude, but nevertheless useful, rule-of-thumb method of trying to assess the mouldability of a material is to quote its *flow path ratio*. This is the maximum ratio of flow length to flow cross-section thickness that may be achieved under normal moulding conditions. The flow path ratio is to some extent dependent on thickness, so that some suppliers will provide data for a range of section thicknesses.

A typical figure for a moderate flow grade of ABS under typical moulding conditions would be 120:1. Thus if the moulding had a thickness of 2.5 mm the distance from the gate to the

PRODUCT DESIGN PRINCIPLES AND PROBLEMS

blind end of the moulding should not be more than 250 mm (about 1 ft).

Whilst it is possible to increase flow path ratios to some extent by raising the melt temperature or increasing the injection pressure, such actions can lead to other problems. For example, raising the temperature may lead to degradation problems (which in the case of ABS can lead to the production of noxious fumes as well as poor mouldings), while use of higher injection pressures may lead to over-packing around the gate with possible problems of injection. It is far better to ensure that the mould design is such that the position of the gate with respect to the furthest blind end should be within limits suggested by the flow path ratio.

Another design aspect associated with flow considerations is the possibility of weld lines. Wherever possible the design should try to avoid split flow paths meeting up again to form a weak weld line (also known as a welding line or a knit line). In some cases this is not possible, but where there is some choice of weld line position, then this should be sited at a point where stresses in service are at a minimum. Some variations in strength at a weld line may be obtained by varying moulding conditions.

Some collected values for maximum flow path ratio are given on page 184 of the data section of this handbook.

FURTHER READING

This chapter has only highlighted certain aspects of design, where possible in terms of the nature of plastics materials. Most suppliers of raw materials, particularly of the so-called engineering thermoplastics, supply excellent data handbooks giving detailed recommend-ations on product design using their materials. Amongst other texts the author has found the following particularly useful:

Beck R. D. (1970). *Plastic Product Design*. New York; Cincinnati; Toronto, London; Melbourne: Van Nostrand Reinhold.

Margolis J. M., ed. (1985). *Engineering Thermoplastics*. New York: Marcel Dekker.

Pye R. G. (1989). *Injection Mould Design*. 4th ed. (Revised). London: Geo. Godwin (now part of the Longman imprint).

PART

THREE

CONTROL

8

STANDARDIZED TESTING

REASONS FOR STANDARDIZED TESTING

Testing procedures are important to suppliers, processors and users for many reasons of which the most important are:

- To ensure that incoming raw materials are of an acceptable and consistent quality,
- To ensure that products of intermediate stages of manufacture are of an acceptable and consistent quality,
- To ensure the end-product of the overall process is of consistent and acceptable quality,
- To evaluate new or competitive materials or modifications to a process,
- To evaluate the fitness for purpose of a material, process or product,
- To obtain early warning of changes that may be taking place in a process even though the product may still be within specification.

The last reason is associated with quality control charts and is dealt with further in Chapter 9.

Failure to test adequately can lead to the use of unsuitable or unnecessarily expensive raw materials, products unacceptable to customers, possible litigation and jeopardization of the long term interests of a company.

STANDARD CONDITIONS

A test is of little value unless there is a reasonable expectation that if it is repeated similar results may be obtained within acceptable limits of experimental error. To realize this expectation, it is therefore important to carry out the tests under standard conditions. Since the results of these tests may form the basis of commercial contract or otherwise used between companies, laboratories and other organizations, it is helpful if the standard conditions are in common use rather than simply confined to a single laboratory. To this end there are a number of national standards organizations such as the American Society for Testing and Materials (ASTM), the British Standards Institution (BSI) (which issues BS standard specifications both for products and methods of test), the German standards organization Deutsches Institut für Normung (DIN) and the Japanese Industrial Standards Committee (JIS). Many tests in these standards are virtually common and some have been published as ISO standards by the International Standards Organization.

Whilst standardized testing has many common features whatever its purpose the approach will be affected by the purpose of the test. If the test is to ensure consistent quality in a raw material, experience may have shown that one or two very simple tests such as measurements of density or refractive index may suffice. On the other hand, the possible application of a plastics material in an aircraft application such as for example a helicopter rotor blade will involve exhaustive testing under a wide range of conditions, initially with simple test pieces; eventually with full-scale products. Many of the tests used for quality control or specification purposes are what are known as *single point tests*. With such tests measurement is made at a single temperature, testing rate with standard sample dimensions, methods of sample preparation and conditioning. For design and product evaluation purposes *multi-point* tests may be used. These tests make measurements under a variety of conditions as deemed appropriate and may include measurements at various rates and temperatures.

Test methods are devised for many purposes and include tests for mechanical properties such as strength, electrical properties, optical properties, flammability, chemical tests, weathering tests and evaluation of toxicity. Certain physical properties are frequently mentioned in the polymer data section of the handbook; a list of standards for a selection of these properties is given at the end of this chapter,

after some comments on measurement of the following key properties:

- impact properties;
- softening point;
- heat resistance;
- flammability;
- water absorption.

EXAMPLE: MEASUREMENT OF TENSILE PROPERTIES

As an example of the importance of standardizing test conditions the measurement of tensile properties will be considered. In this case there are five main testing variables, namely:

1 temperature;
2 straining rate
3 method of specimen preparation;
4 conditioning;
5 size and shape of test piece.

Temperature. It is a frequently, but erroneously, held view that the tensile strength of a thermoplastics material only changes as the softening point is approached. That this is not true can be seen from Figure 8.1 which shows how the tensile strength of the glassy thermoplastic, polystyrene, changes with temperature.

Even greater changes may be expected with many crystalline thermoplastics where melting of the crystal structures can occur over a wide temperature range. Figure 8.2 shows the effect of temperature on Young's modulus (rather than tensile strength) of two types of nylon with substantial changes occurring over a nar-

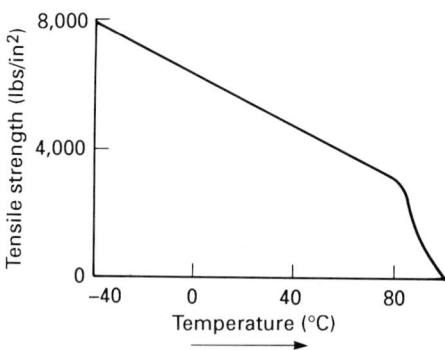

Figure 8.1 *Effect of temperature on the tensile strength of polystyrene*

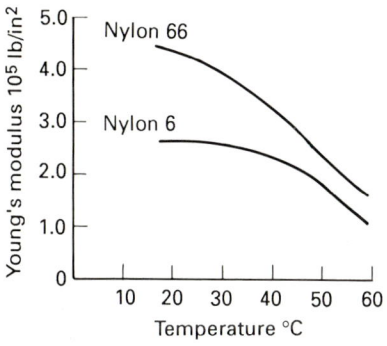

Figure 8.2 *Effect of temperature on the Young's modulus of nylon 66 and nylon 6*

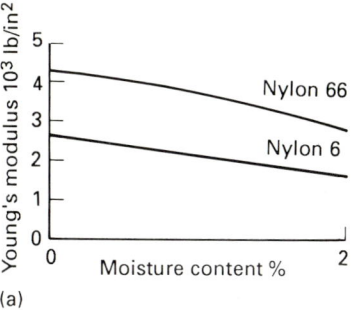

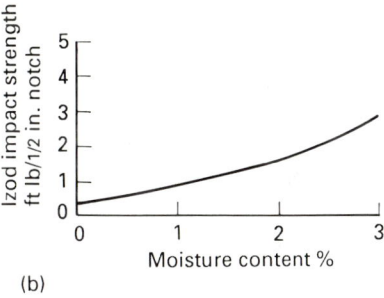

Figure 8.3 *(a) Effect of moisture content on the Young's modulus of nylon 66 and nylon 6. (b) Effect of moisture content on the impact strength of nylon 66*

row temperature range not greatly above normal ambient temperatures. For much reference testing it may be necessary to keep laboratories between closely controlled temperature limits.

Straining rate. This can be particularly critical with crystalline thermoplastics with the results not always predictable. Some years ago the author encountered the data given in Table 8.1 for two grades of polyethylene, each of which had been tested at four different straining rates.

Table 8.1

Straining rate (in/min)	Tensile strength (PSI)	
	Grade A	Grade B
6	2680	1600
12	2750	1580
18	2900	1500
30	3200	1400

Conditioning. Results may well be affected by the previous history of the specimen. For example humidity will have an effect on hygroscopic polymers such as nylon 66. This is shown in Figure 8.3.

In the case of some crystalline thermoplastics, the additional crystallization that occurs after moulding may not only cause *after shrinkage*, but also modify properties such as Young's modulus. In addition, annealing or other heating of injection moulded or extruded specimens may reduce frozen-in orientations and affect the results.

Method of specimen preparation. When polymers are being shaped in the melt, molecular orientation occurs which normally only partially disappears before the melt cools, for example in the injection mould cavity. The amount of such *frozen-in orientation* will depend on the process used being greater in high shear processes, such as injection moulding, and less in low shear processes, such as rotational moulding, and in processes with longer cooling times such as in compression moulding of thermoplastics. When preparing samples by injection moulding, the gate is usually placed at one end of the tensile specimen and orientation is generally along the major axis of the specimen. This will give higher values than would be obtained if the samples were less orientated (as when the samples are compression moulded). Even with a standard moulding with fixed gate positions, it has been found that alteration of the moulding conditions can have a profound effect on many properties, including tensile strength and modulus as shown in Table 8.2.

Table 8.2 Range of Values of Physical Properties of Injection Mouldings Obtained by Alteration of Moulding Conditions (After E. O. Allen and D. A. Van Putte, *Plastics Engineering*, 30, 37 (1974))

Property	Polystyrene	Polypropylene
Tensile load at failure (N)	698–1120	578–898
Flexural strength (MPa)	35–57	16–28
Izod impact strength (ft.lb/in notch)*	1.4–4.0	0.9–8.3
Ball drop impact strength (J)	0.11–6.67	5.2–13.6
Shrinkage across flow (mm/mm)	0–0.005	0.007–0.016
Shrinkage with flow (mm/mm)	0–0.006	0.010–0.014
Stiffness modulus (MPa)	1520–2140	620–1240

* Izod impact data cannot realistically be converted to SI units from the f.p.s. units of the original data.

This immediately causes problems when comparing materials which, because of their nature, have to be processed under different conditions, particularly of temperature.

Sample shape and size: It is common practice to use dumb-bell shaped samples when tensile testing as in Figure 8.4. This is in order that the risk of breaking the sample in the grips (or jaws) of the testing machine are reduced. They should also give a more homogeneous strain in the neck. The size and general dimensions of the sample may also affect the results obtained.

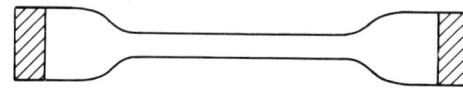

Figure 8.4 *Dumb-bell-shaped tensile test piece. Hatched areas indicate position of grips*

Whilst the above discussion has been largely confined to tensile and other mechanical properties, similar effects may be observed with many other types of test. For example, measures of the electrical properties of power factor and permittivity are greatly affected by temperature and frequency (Figure 8.5).

Widely differing results for softening points may also be obtained depending on the test method used. In Table 8.3 results are given for some amorphous and some crystalline thermoplastics for softening point, as measured using the Vicat test and by the deflection temperature under load test, for which two alternative stressing levels may be used. While amorphous

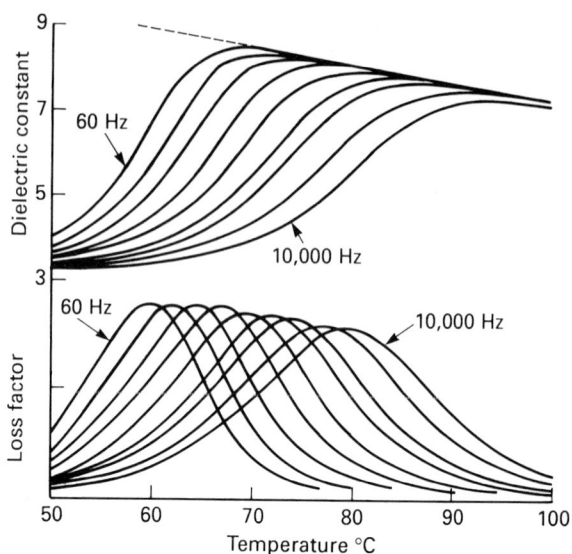

Figure 8.5 *Electrical properties of poly(vinyl acetate) (Gelva 60) at 60, 120, 240, 500, 1000, 2000, 3000, 6000 and 10 000 Hz. (Copyright 1941 by the American Chemical Society and reprinted by permission of the copyright holder.)*

thermoplastics are not greatly affected the results for crystalline materials differ widely. Incorporation of glass-fibre reinforcement into a crystalline polymer reduces these differences considerably.

Table 8.3 Comparison of Polymer Softening Points

Polymer	Deflection temperature under load (°C)		Vicat softening point (°C)
	1.8 MPa	0.48 MPa	10 N 50 °C/hr
Polypropylene	67	127	134
UPVC	64	70	85
PMMA	97	106	114
Polyacetal	100–110	158–170	162–185
Nylon 66	75	200	240
Polycarbonate	127–138	138	155

SOME SELECTED PROPERTIES

IMPACT PROPERTIES

The toughness or impact resistance test of a material is an attempt to assess the resistance of a plastics material or product to a sudden blow. It may be shown that the energy to break is the area under a stress–strain curve up to the point of fracture. This curve will not be the same as that obtained in a slow, e.g. tensile, test but one specific to the stresses and strain rates employed. As a rule, plastics below the glass transition temperature do not extend more than a few per cent before fracture, so that the energy to break will also be low; whereas a polymer which extends several hundred per cent above the T_g (as do many polymers), will have a higher impact energy to break.

It is important to appreciate that a polymer which may appear tough when exposed to a tensile load, may be brittle when assessed by an Izod type test piece, where a notched sample is subjected to a sudden bending load. Table 8.4 is an attempt to summarize the response of various polymers to different stresses.

Table 8.4

Type of stress	Polymers ductile at 25 °C and 1 min^{-1} strain rate
1. Around Izod notch	Low density polyethylene
	Cellulose nitrate and propionate
	ABS and toughened polystyrene
	Polycarbonate (of bisphenol A)
2. Tension	Above materials plus:
	High density polyethylene
	Polypropylene
	Acetal polymers
	Aliphatic polyamides (nylons)
	PPO
	Polyethylene terephthalate
	Polysulphones
3. Simple shear	Above materials plus:
	Polymethyl methacrylate
4. Compression	Above materials plus:
	Polystyrene

For comparison purposes four types of test are deserving of mention. They are:

The Izod test, the most widely used, in which a rectangular bar is clamped at one end. It is then struck at a fixed point above the clamp by a swinging pendulum and the energy required to break the sample assessed from a knowledge of the weight of the pendulum and the amount of follow through of the pendulum after it has broken the sample. Most commonly, the sample is notched to concentrate the stress. Results are influenced considerably by the presence or absence of a notch, the notch radius, whether the notch is machined or moulded-in and the dimensions of the test piece. A schematic arrangement of the test is shown in Figure 8.6.

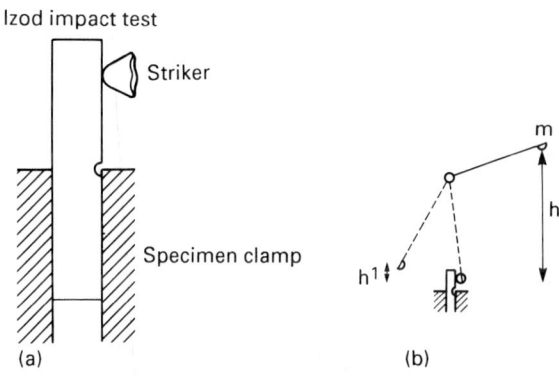

Izod impact test

Striker

Specimen clamp

m

h

h^1

(a)　　　　　　　　　　(b)

Figure 8.6 (a) Arrangement of specimen for Izod test. (b) Diagram showing that impact energy = mg (h–h')

Izod tests are quite reproducible, at least within laboratories, but it is not possible to scale up the results and predict the energy requirements to break a bar of large cross-section from a test carried out with a bar of small cross-section. It should be noted that *highest values are obtained when there is a high level of molecular orientation in the moulding which may be obtained by the use of low melt and mould temperatures.*

A variety of methods are employed for quoting results including the following:

1 Energy to break. This is quoted usually in joules or ft lbf. The result will only apply to one set of sample dimensions and it is not possible to allow for any change in specimen dimensions.
2 Energy to break per unit width of sample. (In notched specimens this is taken as the energy to break per unit width of notch. In this case the results are quoted in such units as ft lb per inch of notch or joules per metre of notch.* The values obtained depend on the notch width. For example, it has been found that reducing the notch width from ½ to ⅛

* 1 ft.lb/in of notch = 53.38 J/m.

141

inch increased the impact strength of a poly-carbonate by a factor of 5.4.)

3 The energy to break per unit fractured area. This will be the cross-sectional area of the sample less the area of the notch (as projected on the cross-section). The units used in this method are most commonly kJ m^{-2}, kgf cm cm^{-2} and ft lbf in^{-2}. These are related by the factors

$$1\,kJ\,m^{-2} = 1.02\,kgf\,cm\,cm^{-2} = 0.476\,ft\,lbf\,in^{-2}.$$

Because Izod impact data on polymers has originated from a variety of sources using different test pieces and methods of quoting results it is often very difficult to compare data.

An alternative to the *Izod test* is the *Charpy method* in which the sample is supported, but not gripped, at each end and subject to impact at the centre. A notch may or may not be present. Correlation between Izod and Charpy test results are quite good and there is little point in providing data from both types of test.

Somewhat different results may be obtained using the *falling weight* test, in which a weight is dropped onto a flat or domed surface. This test requires the use of a great number of samples. Weights are dropped from different heights using several samples for each height in order to assess the energy required to break half the samples at that energy loading. For flat sheet *a high level of uniaxial orientation tends to give low levels of impact resistance: conditions achieved using high melt and mould tempertures. Thus moulding conditions required to give high Izod values may give low falling weight values.* This situation is somewhat different in domed mouldings made by biaxial stretching of sheet since there will be biaxial orientation enhancing the toughness. A fourth method of test is the *tensile impact test* mainly of use in theoretical studies.

In summary, it may be said that for general quality control and for initial comparisons between materials, the Izod test is probably the most suitable. For critical applications which may also involve such aspects as creep fracture much more thorough multi-point and multi-mode testing will be required.

HEAT RESISTANCE AND SOFTENING POINT

Information is frequently required concerning the maximum service temperature at which a polymer may be used. Two quite separate factors need to be considered here:

1 The softening point of the material; and
2 The long term heat stability, i.e. resistance to degradation at elevated temperatures.

Many softening point tests have been evolved over the years and generally these note the temperature at which a sample of material is deformed by an arbitrary extent by an arbitrary load, i.e. when it reaches some arbitrary modulus. This can cause problems, since although with amorphous plastics there is a fairly sharp drop in modulus around the glass transition temperature, in crystalline polymers crystal melting occurs over a wide temperature range, through which there is a steady change in modulus (see Figure 8.7). Two tests now dominate softening point assessment:

1 The deflection temperature under load test (also known as the heat distortion temperature test).
2 The Vicat test.

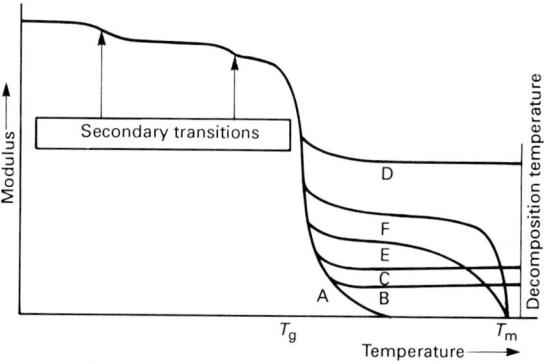

Figure 8.7 *Schematic illustration of dependence of the modulus of a polymer on a variety of factors. A is an amorphous polymer of moderate molecular weight whereas B is of such a high molecular weight that entanglements inhibit flow. Similar effects are shown in C and D, where the polymer is respectively lightly and highly cross linked. In E and F the polymer is capable of crystallization, F being more highly crystalline than E or containing fibre reinforcement. From Brydson, Plastics Materials, Butterworth, 5th ed., p. 177, Figure 9.1*

In the deflection temperature test, the temperature is noted at which a bar of material, subjected to a three point bending stress, is deformed by a specified amount. In the standard tests, a load is applied which produces a maximum stress at the mid-point of the beam of either 1.82 MPa (264 p.s.i.) or 0.48 MPa (66 p.s.i.). For the reasons given above, the softening points obtained using the two loadings are similar for amorphous materials but different for unfilled crystalline polymers.

In the Vicat test, the temperature is noted at which a needle of cross-sectional area of 1 mm^2 indents the sample by 1 mm under a specified load (usually 10 N). Vicat results are similar to 66 p.s.i.-deflection temperature test results. (See Table 8.3).

A knowledge of *long term heat stability* is required when considering plastics for many applications. Raw material suppliers are increasingly quoting the *UL Temperature Index* assessed by methods devised by the Underwriters Laboratories. In these tests a large number of samples are aged in ovens set at a range of temperatures over a period of time (e.g. a year). These samples may be test pieces for testing one or more of such properties:

- tensile strength
- flexural strength
- Izod impact
- tensile impact
- electrical strength.

Samples for the appropriate tests are withdrawn at intervals and tested and compared with the original control value. A plot is then made of the percentage retention in the value of a property (cf. the original control value) against time and the time noted to give a 50% reduction. This is quite arbitrarily known as the failure time. This procedure is repeated for ageing tests over a range of temperatures. (See Figure 8.8(i).)

The higher the temperature the shorter the failure time. The UL Temperature Index, for a particular property, is taken, again quite arbitrarily, as the temperature at which it takes 10,000 h to fail a particular property (i.e. to drop in value by 50%). This is determined by means of a plot of log (failure time) versus 1/K where K is the temperature in degrees Kelvin. (See Figure 8.8(ii).)

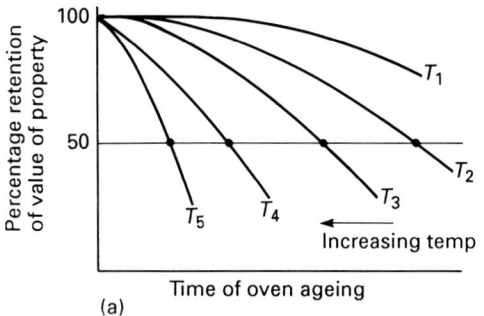

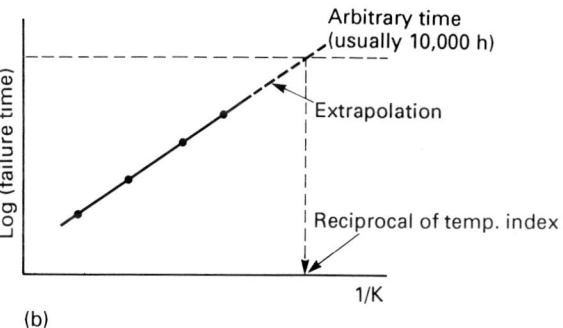

Figure 8.8 *Method for determining the UL temperature index*

Since different properties will change at different rates and because results may also depend on the thickness of the specimen, *there is no single value for temperature index but rather a matrix of values*. Most commonly, 'Mechanical, without impact' data is quoted, based on tensile test results.

FLAMMABILITY

For many applications of plastics considerations of flammability are particularly important. This is especially the case where plastics are being used in situations where escape is difficult, for example submarines, aircraft and underground. It is also a problem in domestic situations where there is frequently a lack of fire fighting equipment and knowledge, but where potentially lethal materials are present.

In order to assess suitability for a purpose, it may be necessary to test the fire resistance of the structure, be it a mock up of a building or a sample piece of furniture. In addition many tests have been developed to try and rate plastics materials in terms of their flammability. Two tests are now very widely used for this purpose:

1 The Limiting Oxygen Index Test, and
2 The Underwriters Laboratories UL94 ratings.

The test method and data for these properties are given in the data section of this handbook and will not be considered further here.

It is to be noted that most victims of fires die through smoke inhalation causing suffocation or poisoning. As a test for assessing smoke density the NBS Smoke Chamber Test has become widely adopted.

WATER ABSORPTION

Water absorption is not only of importance to the processor because of problems of handling damp materials, but also for the designer and user of plastics products. The presence of water in a moulding can have the following effects:

- It will cause the product to swell;
- It may plasticize the polymer leading to a reduction in stiffness (but possibly an increase in toughness);
- It may act as a carrier allowing the polymer to absorb bacteria, undesirable chemicals and stains.

Relevant tests should be carried out to check all of these aspects.

The most common test is to measure water absorption by measuring the change of weight of an exposed polymer specimen. Results may be expressed in a number of ways. In many tests results are simply expressed in terms of mg water absorption and it must be stressed that such results will depend considerably on the size and shape of the test piece, the time of immersion and the temperature of immersion. While expressing the results as a percentage may seem more meaningful, results will also be affected by shape, size, time and temperature.

SOME STANDARD TESTS

The following table lists some standard tests used for comparison of materials. The ISO, ASTM, BS and DIN tests may differ in detail.

Table 8.5

Property	ISO	ASTM	BS	DIN
Izod impact	DIS180	D256	2782 (Method 305A)	–
Charpy impact	DIS179	D256	2782 (Method 351A)	53453
Tensile impact		D1822	–	53448
Tensile strength	R527, R1184	D638	2782 (Methods 320, 326 and 1003)	53455
Vicat softening pt	306	D1525	2782 (Method 120)	53460
Deflection temperature	75	D648	2782 (Method 121)	53461
Limiting oxygen index		D2863	2782 (Method 141B)	–
Density	R1183	D792	2782 (Methods 620 A to E)	–
		D1505		
		D1622		
Hardness (Shore)	R868	D2240	2782 (Method 365B)	53505
Volume resistivity	R1325	D257	2782 (Method 230A)	53482

STATISTICS AND QUALITY CONTROL

INTRODUCTION

Successful quality control of a product requires not only that the product meets its specification, but in addition, monitoring of the process to ensure that adverse changes are not occurring, even though the product is still within specification. If this is not done a situation may arise where within a short space of time high reject rates may develop. Such problems may be anticipated by the use of statistical quality control methods. Furthermore, such methods may be useful in avoiding acceptance of specification limits that may be difficult to achieve.

This chapter commences with a brief introduction to some basic statistical concepts and then demonstrates how these may be used for statistical quality control methods. Clearly the subject of statistics and statistical quality control can only be touched on here and for more information the reader is referred to more de-

tailed texts. One the author has found very useful is *Basic Statistical Methods for Engineers and Scientists* by A. M. Neville and J. B. Kennedy, Intertext, 1984. For quality control charts and related topics *Statistical Process Control* (2nd edn, 1990) by J. S. Oakland and R. F. Followell, Heinemann Professional Publishing, Oxford, is also particularly recommended.

SCATTER OF EXPERIMENTAL RESULTS

If a series of mouldings taken from a single mould cavity are weighed on a sufficiently accurate balance it will be observed that they will vary in weight. If the number of hours of sunshine in a day is measured at a particular site for a month, different values will be noticed for each day. If a group of students are asked to measure individually the perimeter of a classroom using a metre rule to the nearest millimetre, it is also expected that they will obtain different results. Such differences are referred to as a *scatter* (or spread) of results and are common in any experimental work. In the first two examples, the differences will be real, whereas in the third example it may be assumed that the room has not changed its size and that the difference in results is due to *experimental error*. Providing the experimental errors are *random* (and not due to incorrect calibration of the measuring instrument which would give *biased* results) we can treat all the three types of scatter in the same way.

As an example, let us consider the number of rejects in a series of moulding shifts. Let the number of rejects per shift in six successive shifts be

$$5 \quad 5 \quad 7 \quad 3 \quad 4 \quad 6.$$

The number of rejects is an example of a variable whose individual values are denoted in this chapter by the symbol X.

If the number of measurements of the variable (in this case 6) is denoted by N, then the *arithmetic mean* or *average* is given by

$$\bar{X} = \frac{\Sigma X}{N}$$

where Σ denotes the sum of all the values of X.

Thus the average number of rejects in the above example will be $30/6 = 5$. It will also be noted that the average difference from the mean is equal to 1 (we say that the mean deviation = 1). In analysing results statistically it is more useful to use the *root mean square deviation*, better known as the *standard deviation*. This is given by the equation

$$\sigma = \sqrt{\frac{\Sigma (x - \bar{x})^2}{N}}$$

In this instance $\sigma \simeq 1.291$

The above formula can be shown to give biased results with small samples, where it is more accurate to use the equation

$$s = \sqrt{\frac{\Sigma (x - \bar{x})^2}{N-1}}$$

where s is known as the *best estimate of the standard deviation obtained from a sample* or often simply as the *sample standard deviation*. As the sample size increases so the values of σ and s converge. In the above example $s \simeq 1.414$.

Whilst the computation of standard deviations has been facilitated in recent years by computers and by pocket calculators which are programmed for rapid calculation of σ and s, it is also sometimes of use to estimate the standard deviation from the range (R). This is given by the formula

$$R = \text{Highest score} - \text{Lowest score} + 1.$$

If we have a number of samples each with its own value of the range (R), then an estimate of the standard deviation can be obtained from the mean value of the sample range $\bar{R}$ by the formula

$$s = \bar{R}d.$$

The coefficient d varies with N, some examples being given in Table 9.1.

Table 9.1

N	d	N	d	N	d
2	0.8862	6	0.3945	10	0.3249
3	0.5908	7	0.3698	12	0.3069
4	0.4857	8	0.3512	15	0.2880
5	0.4299	9	0.3367	20	0.2677

The range is an *inefficient* measure of scatter in that only some of the results (the two extreme values) are used in computing its value, but where the sample is small then less information is discarded. For example; with a sample of 4 the range will be 50% efficient.

The way in which a set of individual results are scattered is known as a *distribution*. It has been observed that many experimental results are distributed in a manner that with a large number of samples resembles one of two theoretical distributions, namely the Gaussian and Poisson distributions. The *Gaussian distribution*, sometimes known as the *Normal distribution*, has a shape of general form shown in Figure 9.1.

The vertical axis is a measure of the probability that an individual result will have the value

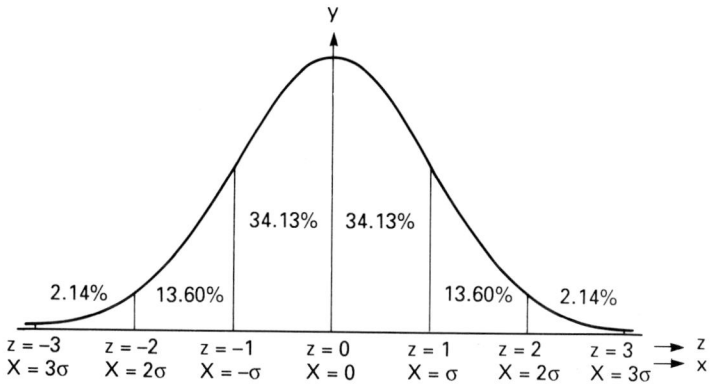

Figure 9.1 *The normal probability curve (Gaussian distribution) showing the percentage of results within 1, 2 and 3 standard deviations of the mean. In this diagram the mean is set at zero and z = x/σ*

of the variable X at that point. It will be noted that results centre and tend to concentrate around the mean and are less frequent the further they are away from the mean. Whilst there are some variables for which a distribution of individual results does not fit a Gaussian distribution, a distribution of sample means of a variable almost invariably does so.

It may be shown that in a Gaussian distribution, 95% of the results lie in the range $\overline{X} \pm 1.96\sigma$ (i.e. within about 2 standard deviations of the mean) and that 99.8% of the results lie within $\overline{X} \pm 3.09\sigma$ (i.e. about 1 in 1000 results will be more than $\overline{X} + 3\sigma$) (Figure 9.2).

If a very large number of samples of size N are taken from such a distribution, then it may further be shown that 95% of the sample means will lie in the range

$$\overline{X} \pm 1.96\sigma/\sqrt{N}$$

and that 99.8% will be in the range

$$\overline{X} \pm 3.09\sigma/\sqrt{N}.$$

CONFIDENCE LIMITS

The above expressions are known respectively as the 95% and 99.8% *confidence limits*. These mean that *if the standard deviation is known accurately* one may be 95% confident that the true value (or true mean value) will lie within $1.96\sigma/\sqrt{N}$ of the measured sample mean.

In practice we have to estimate σ from the sample standard deviation s and the Gaussian distribution has to be replaced by a series of distributions whose shape depends on the sample size. These are known as *t*-distributions. Some *t*-distribution data is given in Table 9.2.

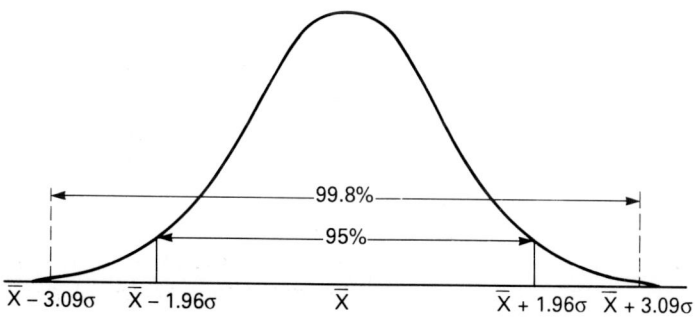

Figure 9.2 *Confidence limits of a Gaussian distribution commonly used for quality control purposes*

Table 9.2 Some selected values for *t*

Degrees of freedom	Probability		
	0.05	0.01	0.001
1	12.71	63.66	636.62
2	4.30	9.93	31.60
3	3.18	5.84	12.94
4	2.78	4.60	8.61
5	2.57	4.03	6.86
6	2.45	3.71	5.96
8	2.31	3.36	5.04
10	2.23	3.17	4.59
15	2.13	2.95	4.07
25	2.06	2.79	3.73
60	2.00	2.66	3.46
∞	1.96	2.58	3.29

In this case the *95% confidence limit* will be given by $\bar{X} \pm t_{0.05}s/\sqrt{N}$ and the *99.8% confidence limit* will be

$$\bar{X} \pm t_{0.002}s/\sqrt{N}.$$

(The suffixes refer to the fraction of results outside the range.)

The values of $t_{0.05}$, $t_{0.002}$ etc will depend on N (or more correctly on what are known as the *degrees of freedom* which in this case are given by $N-1$).

Example:

Five samples of a polymer are measured for hardness. The results give a mean value of 62 and a standard deviation of 2.

Assuming that our sampling method is unbiased we can have a 95% confidence that the 'true' value for hardness is in the range 62 ± 2.78 (2/5), i.e. between 59.51 and 64.49.

PARETO ANALYSIS

Pareto analysis is based on a very empirical relationship first noted by the Italian economist in the nineteenth century when he observed that 80–90% of Italy's wealth lay in the hands of some 10–20% of the population. Since then similar relationships have been observed in many other fields of study including quality. For example, it is often found that injection moulding rejects may be due to a large number of causes, but that the bulk of the rejects may be due to just a small percentage (e.g. 2–3) of these.

It is clearly worthwhile to pay attention to such major causes of defect since if these can be removed then the failure rates may be much reduced. A typical Pareto curve for injection mouldings is shown in Figure 9.3.

This has been obtained by analysing the defects, calculating the percentage of defectives

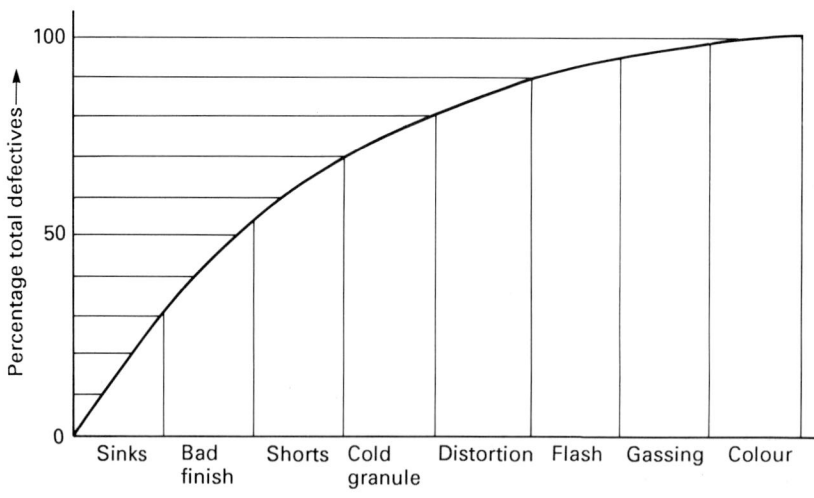

Figure 9.3 *Typical Pareto curve for injection moulding defects*

from each fault, listing the faults in order of frequency along the base line of a graph (giving the same distance along the base line for each defect) and then plotting on the vertical axis at a point corresponding to the right-hand side of each vertical column the percentage total defectives for all the defects to the left of that line.

Thus in the figure it is seen that about 30% of the defects are due to sink marks, about 55% due to sink marks and bad finish combined and 70% due to sinks, bad finish and shorts. Elimination of these three defects will thus reduce the defects by 70%. In this particular instance all could be due to a low injection rate and/or low packing pressure.

In some processes the presence of one defect may be more costly than another. For example a gross contamination of a length of extrudate by metal may mean that the whole batch be scrapped; the presence of a die line along the length may mean that the material will have to be reworked, while some flow problems at the end of a run due to clogging up of the breaker plate may mean the loss of only a small part of the production run. In such circumstances it may be more useful to plot a Pareto curve, not of cumulative percentage defectives but rather cumulative percentage costs of the defects.

Pareto diagrams have become extensively used to identify important problems in maintaining quality and establishing priorities for action. Whilst it helps to ensure that the maximum reward is obtained for the effort expelled, it does not mean that correction of a less frequent fault which can be effected by some simple and obvious adjustment should be delayed until higher ranked defects are tackled.

SPECIFICATION AND TOLERANCE LIMITS

The manufacture of a product commonly involves meeting *specification limits*. For example, the product may be a ring and one of the requirements is that the inside diameter will be within the limits 3.825 ± 0.005 cm. By the nature of the manufacturing process there will be a scatter in the values of the inside diameters of the rings. If for a moment we assume that we know the mean and standard deviation accu-

rately and that the scatter is Gaussian, then it may be shown that

50% of the results are within 0.6745σ of the mean
68.26% are within 1σ of the mean
95% are within 1.96σ of the mean
95.44% are within 2σ of the mean
99.74% are within 3σ of the mean
99.8% are within 3.09σ of the mean.

These limits are a direct consequence of the properties of the Gaussian distribution and are often referred to as *natural tolerance limits*.

If the average ring diameter was the same as the central value of the specification limit (3.825 cm) and the standard deviation were only 0.001 cm, then 99.74% of the rings would be within the range 3.825 ± 0.003, i.e. well within the specification and the number of rejects expected would be very low. If, on the other hand, it was found that the standard deviation was 0.005 cm (with the mean unchanged), 31.74% of the rings would be expected to be outside the specification limits.

In this second example the manufacturer has three possible choices:

1 To accept a high reject rate.
2 To persuade the customer, or whoever lays down the specification limits, to accept wider limits.
3 To improve the accuracy of the process.

Conversely it might be argued that in the first example the process was 'too good' and that it might be possible to use a somewhat less costly procedure.

As a general rule, providing the average figure coincides with the central value of the specification and that the + and − tolerances are equal, it is useful to aim for specification limits to range slightly more than 6σ.

QUALITY CONTROL CHARTS

The progress of a manufacturing process may be followed by the use of *quality control charts* which make use of the concept of tolerance limits. The purpose is to check that the products are varying in line with that expected from Gaussian scatter, in which case the process is said to be *stable* or *in control*. The visual presentation in chart form enables changes in the process to be quickly identified and remedial action taken, hopefully before there is any adverse influence on the failure rate. In the following section an outline will be given of the procedure for making a control chart and in the subsequent section some examples will be given concerning interpretation. At this stage it will be assumed that the property being measured has a numerical value, i.e. it is a *variable* as opposed to an unquantified attribute, e.g. go–no go, pass–fail.

A control chart provides a visual way of observing how individual sample results compare with the natural tolerance limits. It is common to prepare two charts, *a mean chart* and *a range chart*. A typical chart for means is shown in Figure 9.4. To construct such a chart it is first necessary to estimate the mean value of the property under consideration from sample means taken over a period of time.

We then calculate what are known as the *inner or warning limits*. These are usually based on the Gaussian distribution property that 95% of the results are within the range

$$\bar{X} \pm 1.96s/\sqrt{N},$$

i.e. 2.5% will be above the upper inner limit and 2.5% below the lower inner limit. In practice it is usual to estimate the value of s from the mean range using the equation

$$s = \bar{R}d$$

Hence the values for the mean warning limits (MWL) will be given by

$$MWL = \bar{\bar{X}} \pm 1.96\bar{R}d/\sqrt{N}$$

$$= \bar{\bar{X}} \pm A_w\bar{R},$$

where $A_w = 1.96d/\sqrt{N}$ and $\bar{\bar{X}}$ is the grand mean of the sample means. Some figures for A_w are given in Table 9.3.

Table 9.3 Warning and Action Factors for Mean Chart

Sample size (N)	2	3	4	5	6	7	8	9	10	11	12
A_W		1.23	0.67	0.48	0.38	0.32	0.27	0.24	0.22	0.20	0.19	0.17
A_A		1.94	1.05	0.75	0.59	0.50	0.43	0.38	0.35	0.32	0.29	0.27

In a similar way we also calculate the *mean action limits*, these being the tolerance values for which only 1 in 1000 results are above the upper

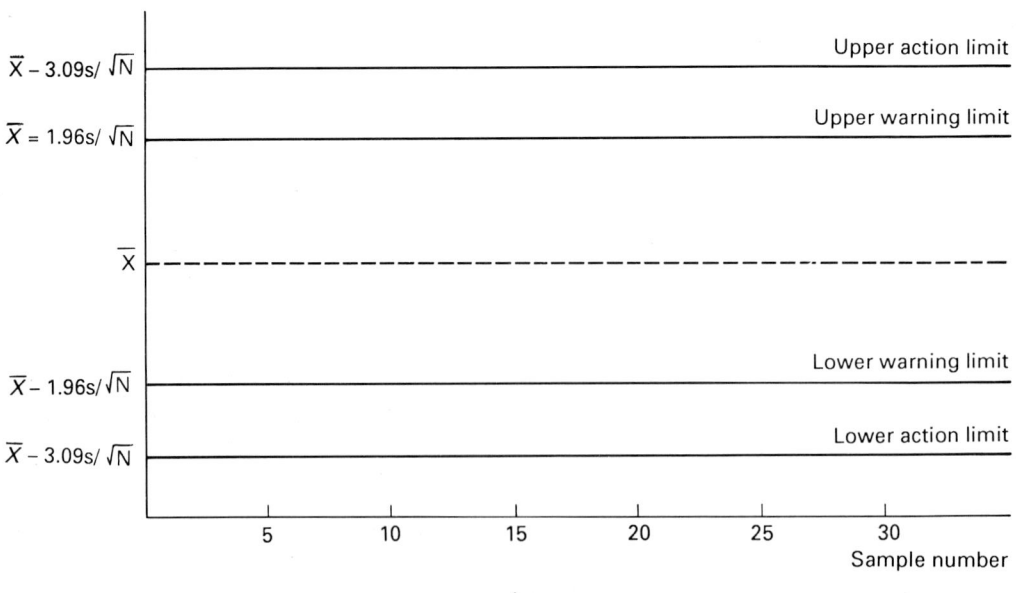

Figure 9.4 *Quality control chart for means – general format*

action limit and 1 in 1000 below the lower action limit. These limits are given by

$$\text{MAL} = \overline{\overline{X}} \pm A_A \overline{R},$$

where

$$A_A = 3.09d/\sqrt{N}$$

Having constructed our mean chart the results of routine testing are recorded on the chart. In Figure 9.5(a) just one result is outside a warning limit and there is no need to take particular action. However in Figure 9.5(b) there is a result outside the action limit, an event only to be expected on 1 in 1000 occasions and it is necessary to investigate (even though the sample may be within specification and not a reject), since something is clearly happening to the process.

It also takes little more than casual inspection of Figure 9.5(c) to see that there is a steady drift in the values of the variable being measured; something is changing in the process and action should be taken before high reject rates develop. In Figure 9.5(d) it is possible to see that the points marked by o are different from those marked by ●. This might be due, say, to the use of a different machine, a different operator or a different mould cavity.

In addition to a mean chart it is also useful to produce a chart which displays scatter of results in a sample and in practice the range chart is mostly commonly used. In this case the *range warning limits* (RWL) are given by

$$\text{RWL} = D_w \overline{R}$$

and the range action limits (RAL) by

$$\text{RAL} = D_A \overline{R}.$$

Some values for D_w and D_A are given in Table 9.4

Table 9.4 Control Chart Limits for Range

Sample size (N)	2	3	4	5	6	8	10	12
D_w (upper)	2.81	2.17	1.93	1.81	1.72	1.62	1.56	1.51
D_w (lower)	0.04	0.18	0.29	0.37	0.42	0.50	0.54	0.58
D_A (upper)	4.12	2.99	2.58	2.36	2.22	2.04	1.94	1.87
D_A (lower)	0.00	0.04	0.10	0.16	0.21	0.29	0.35	0.40

A typical range control chart is shown in Figure 9.6.

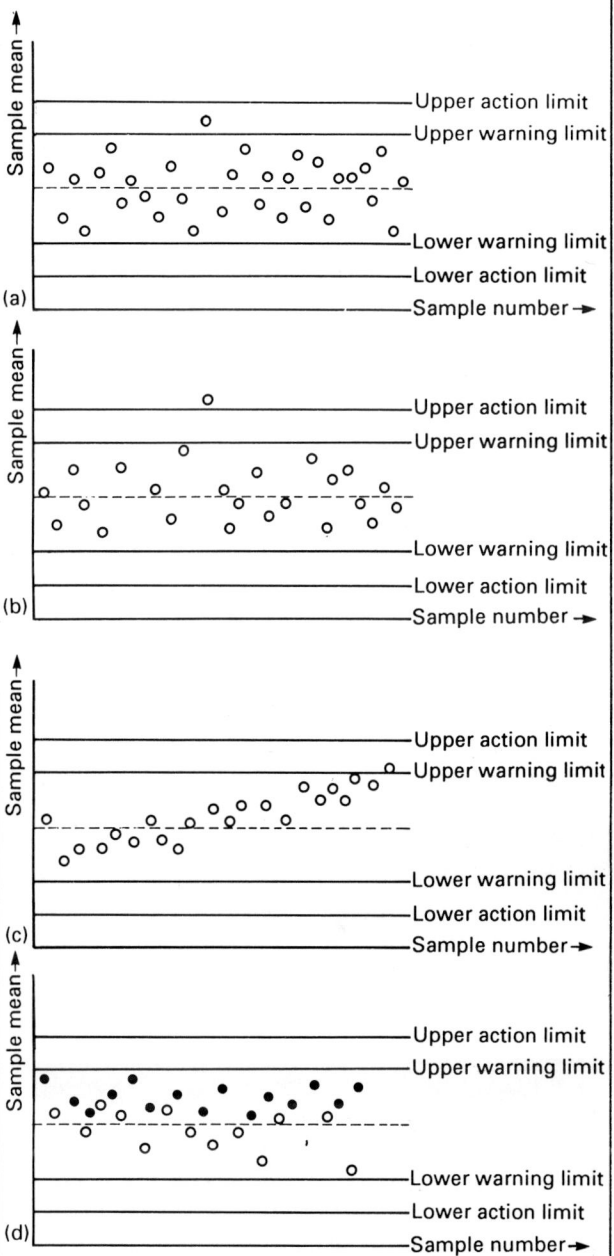

Figure 9.5 (a) One result outside warning limit. (b) One result outside action limit. (c) Drift. (d) Distribution of ● and o points are different

CONTROL CHARTS FOR ATTRIBUTES

On occasions, routine control involves checking whether or not items are defective, or how many defects there may be in an item. In such cases the limits are calculated in a somewhat different way. There is not space to go into this here, except to note that if we are concerned

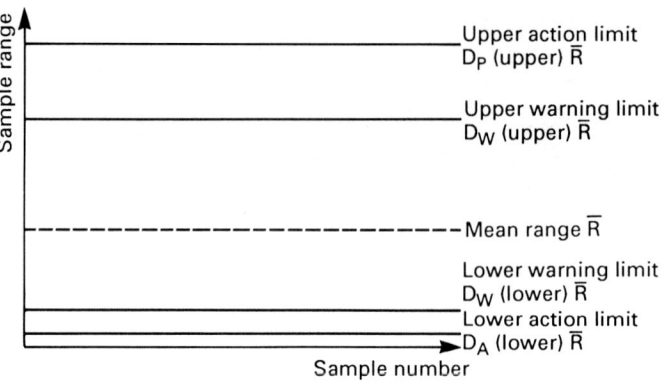

Figure 9.6 *Format of a range control chart*

with the number of items defective in a batch, the MAL for the number of defectives will be given by

$$MAL = N_p \pm 3.09N_p(1 - p),$$

where p is the mean proportion of defective items in a sample of size N. If we are concerned with the number of defects in an item then the limits will be given by

$$MAL = N_p \pm 3.09 \, N_p.$$

QUALITY ASSURANCE

Purchasing and other agreements increasingly involve assurance as to quality. This is reflected in the emergence of the ISO standards 9000–9004 and the British Standard BS 5750 Parts 0–6 which is based upon them. These standards are concerned with *quality systems* which are the organizational structures, responsibilities, procedures, processes and resources for implementing quality management.

In turn *quality management* is defined in BS 5750 as, that aspect of the overall management function that determines and implements quality policy.

Also defined in the standards are *quality control* as the operational techniques and activities that are used to fulfil requirements for quality; and *quality assurance* as all those planned and systematic actions necessary to provide adequate confidence that a product or service will satisfy the given requirements for quality.

The standards provide models for quality assurance in production, installation, in final inspection and testing with guidance on implementation. The quality system elements are grouped into three distinct models:

ISO 9001 – for use when conformance to specified requirements is to be assured by the supplier during several stages which may include design/development, production, installation and servicing.

ISO 9002 – for use when conformance to specified requirements is to be assured by the supplier during production and installation.

ISO 9003 – for use when conformance to specified requirements is to be assured by the supplier solely at final inspection and test.

USEFUL DATA

STANDARDIZED UNITS OF MEASUREMENT AND THEIR CONVERSION

SYSTEMS OF MEASUREMENT

There are occasions where instructions to add a matchboxful of *A* to a sack of *B* will be sufficiently precise for the situation in question. Much more often it is necessary to define quantities in a clear and unambiguous way and this requires the use of standardized units. Such standardized units will be of greatest value when they are internationally recognized.

At the present time there are three international systems in widespread use. These are:

1 The foot-pound-second (f.p.s. system) still widely used in the English-speaking world;

2 The centimetre-gramme-second (c.g.s.) which largely resulted from the encouragement of the French statesman Talleyrand after the French Revolution, and;

3 The International System of Units (Système International d'Unités or SI Units) now adopted by the International Organization for Standardization (ISO) and which is expected to progressively replace the older systems.

It is still necessary to be familiar with the three systems for a number of reasons. Firstly, many of the older sources of information provide data in non-SI units. Second, many of those brought up to use the older units tend to continue to use such older units when possible. The writer is not alone in finding it easier to think in such terms as pounds and pounds per square inch rather than in Newtons and Pascals. A third complication is that in some countries, including Britain, older systems remain in official use, at least in part. For example, in Britain distances are still normally measured in miles.

The rest of this section will be concerned with a description of the SI system together with conversion factors to the other systems.

THE SI SYSTEM OF UNITS

The SI system is built up from seven *base units*. These are given in the following table.

SI Base Quantities and Units

Base quantity	Dimension	Base unit	Symbol
Length	L	metre	m
Mass	M	kilogram	kg
Time	T	second	s
Electric current	I	ampere	A
Temperature	θ	kelvin	K
Luminous intensity	J	candela	cd
Amount of substance	1	mole	mol

The base units are accurately defined. For example a metre is defined as the length equal to 1 650 763.73 wavelengths in vacuum of the radiation corresponding to the transition between the levels $2p_{10}$ and $5d_5$ of the krypton–86 atom. For definitions of other units the reader is referred to standard texts.

In addition to the basic units there are two *supplementary units*, the radian and the steradian. From the base and supplementary units are a variety of *derived units*. Some of the more important derived units are given below.

SI Supplementary and Derived Quantities and Units

Quantity	Dimension	Unit	Symbol	Equivalent
Plane angle	1	radian	rad	$(=180°/\pi)$
Solid angle	1	steradian	sr	

Density	ML^{-3}			$kg m^{-3}$	
Velocity	LT^{-1}			$m\ s^{-1}$	
Acceleration	LT^{-2}			$m\ s^{-2}$	
Momentum	MLT^{-1}			$kg\ m\ s^{-1}$	
Moment of inertia	ML^2			$kg\ m^2$	
Force	MLT^{-2}	newton	N		
Pressure, stress	$ML^{-1}T^{-2}$	pascal	Pa	$N\ m^{-2}$	
Energy, work	ML^2T^{-2}	joule	J	$N\ m$	
Power	ML^2T^{-3}	watt	W	$J\ s^{-1}$	
Viscosity	$ML^{-1}T^{-1}$		$Ns\ m^{-2}$	$kg\ m^{-1}\ s^{-1}$	
Frequency	T^{-1}	hertz	Hz	s^{-1}	
Electric charge	TI	coulomb	C	$A\ s$	
Electric potential	$L^2MT^{-3}I^{-1}$	volt	V	$W\ A^{-1}$	
Capacitance	$T^4I^2L^{-2}M^{-1}$	farad	F	$C\ V^{-1}$	
Resistance	$L^2MT^{-3}I^{-2}$	ohm	Ω	$W\ a^{-2}$	
Magnetic flux	$L^2MT^{-2}I^{-1}$	weber	Wb	$V\ s$	
Magnetic induction	$MT^{-2}I^{-1}$	tesla	T	$V\ s\ m^{-2}$	
Inductance	$L^2MT^{-2}I^{-2}$	henry	H	$V\ s\ A^{-1}$	
Luminous flux	J	lumen	1m	$cd\ sr$	
Illumination	JL^{-2}	lux	1x	$1m\ m^{-2}$	

SI PREFIXES AND MULTIPLICATION FACTORS

The names of the multiples and submultiples of the units are formed by means of the prefixes given in the following table.

Factor by which the unit is multiplied	Prefix	Symbol
1 000 000 000 000= 10^{12}	tera	T
1 000 000 000= 10^9	giga	G
1 000 000= 10^6	mega	M
1 000= 10^3	kilo	k
100= 10^2	hecto	h
10= 10^1	deca	da
0.1= 10^{-1}	deci	d
0.01= 10^{-2}	centi	c
0.001= 10^{-3}	milli	m
0.000 001= 10^{-6}	micro	μ
0.000 000 001= 10^{-9}	nano	n
0.000 000 000 001= 10^{-12}	pico	p
0.000 000 000 000 001= 10^{-15}	femto	f
0.000 000 000 000 000 001= 10^{-18}	atto	a

Note 1. Through common usage certain multiples and submultiples of the base units have been given names although they are not recognized SI units. These include:

micron (μm) $= 10^{-6}$ m
Ångstrom (Å) $= 10^{-10}$ m
tonne (t) $= 10^6$ g $= 100$ kg
minute (min) $= 60$ s
hour (h) $= 3600$ s
day (d) $= 86\ 400$ s
year (a) $\simeq 3.1557 \times 10^7$ s.

Note 2. It should be noted that masses are still expressed as multiples of the gram although the base unit is the kilogram. Thus 10^{-6} kg should be written as 1 mg.

Note 3. Temperatures are frequently expressed in degrees Celsius (or centigrade) or in degress Fahrenheit. The relationship between these and the kelvin is

$$x\ ^\circ C = (x + 273.15)\ K$$
$$= (1.8x + 32)\ ^\circ F$$

In terms of the temperature interval

$$1\ ^\circ C = 1\ K = 1.8\ ^\circ F.$$

Temperature interval is frequently differentiated from actual temperatures in the Celsius and Fahrenheit systems by writing them as deg C and deg F respectively.

SOME USEFUL CONVERSION FACTORS

Quantity	SI Unit	c.g.s. unit	f.p.s. or other
Length	1 m	100 cm	3.281 ft = 39.37 in
Time	1 s	1 s	1 s
Mass	1 kg	1000 g	2.205 lb
Area	1 m^2	10^4 cm^2	10.764 ft^2 = 1550.0 in^2
Volume	1 m^3	10^6 cm^3	35.31 ft^3 = 6.102 $\times$ 10^4 in^3
Frequency	1 Hz	1 c/s	1 c.p.s.
Rotational frequency	1 s^{-1}	1 rev/s	60 r.p.m
Density	1 Mg/m^3	1 g/cm^3	62.43 lb/ft^3 = = 0.03613 lb/in^3
Velocity	1 m/s	100 cm/s	3.281 ft/s = 39.37in/s
Fluid velocity	1 m^3/s	6 $\times$ 10^4 litres/min = 10^6 cm^3/s	2119 ft^3/min = 6.1 $\times$ 10^4 in^3/s
Force	1 N	0.102 kgf = 10^5 dynes	0.2248 lbf = 7.233 pdl
Surface tension	1 mN/m	1 dyne/cm	
Pressure) *	1 MPa	10.20 kgf/cm^2 = 10^7 dyne/cm^2	145 p.s.i.
Stress	1 kPa	10.20 gf/cm^2 = 10^4 dyne/cm^2	0.1450 p.s.i.
Strength	1 Pa	7.5 $\times$ 10^{-3} mmHg	
Modulus	1 mN/m^2	7.5 $\times$ 10^{-6} mmHg	
Torque	1 Nm	10.20 kgf cm	141.6 oz in
Energy†	1 J	1.0197 $\times$ 10^4 gf cm = 0.239 cal = 0.239 cal = 10^7 ergs	0.738 ft lbf = 9.48 $\times$ 10^{-4} Btu
	1 MJ	0.278 kWh = 239 kcal	948 Btu
Power	1 W	0.860 kcal/h	0.00134 HP = 3.41 Btu/h
Momentum	1 kg m s^{-1}	10^5 g cm s^{-1}	7.2307 lb ft s^{-1}
Moment of inertia	1 kg/m^2	10^7 g cm^2	23.730 lb ft^2
Thermal conductivity	1 W/mK	0.00239 cal/cm $^\circ$C s	6.93 Btu in/ft^2 $^\circ$F h
viscosity – dynamic	1 Ns/m^2	10 poise	0.02089 lbf s ft^2
viscosity – kinematic	1 m^2/s	10^4 stokes	
permeability	1 m^4/Ns	1.013 $\times$ 10^9 cm^2/At s	
electrical resistivity	1 Ωm	100 Ωcm	

* Pressure may also be expressed in bars (b)

1 bar = 10^5 N/m^2 = 14.504 p.s.i.

† Electrical, mechanical or heat energy

TEMPERATURE CONVERSIONS AND SATURATED STEAM PRESSURES

The basic SI unit of temperature is the *degree Kelvin*. In practice however temperature difference is normally expressed in *degrees Celsius*, at one time known as degrees Centigrade. The respective symbols for the two types of degree are K and °C.

The units of Kelvin and Celsius degree interval are identical. For practical purposes

$$C = K - 273.15$$

with 0 °C the freezing point and 100 °C the boiling point of water at STP.

The Fahrenheit scale is such that the freezing point of water is 32 °F and the boiling point 212° F. It follows that the Fahrenheit degree interval is 5/9th of the Kelvin and Celsius degree intervals and that

$$F = 1.8C + 32.$$

For temperatures above 100 °C the table gives the temperature equivalents of saturated steam at sea level. The unit conversions are 1 MPa (1 MN/m² = 10.20 kgf/cm² = 145 p.s.i.).

Temperature Conversions up to 100 °C

Celsius	Fahrenheit
−273.15	−459.67
−100	−148
−90	−130
−80	−112
−70	−94
−60	−76
−50	−58
−40	−40
−30	−22
−20	−4
−10	14
0	32
5	41
10	50
15	59
20	68
25	77
30	86
35	95
40	104
45	113
50	122
55	131
60	140
65	149
70	158
75	167
80	176
85	185
90	194
95	203
100	212

Temperature Equivalents of Saturated Steam at Sea Level

Gauge pressure			Approximate temperature	
p.s.i.	kg/cm^2	MPa	°C	°F
0	0	0	100	212
5	0.35	0.034	109	227
10	0.70	0.069	115	239
15	1.06	0.104	121	250
20	1.41	0.138	125	258
22	1.55	0.152	127	261
24	1.69	0.166	129	265
26	1.83	0.179	131	268
28	1.97	0.193	133	271
30	2.11	0.206	134	274
32	2.25	0.221	136	277
34	2.39	0.234	138	280
36	2.53	0.248	139	282
38	2.67	0.261	140	285
40	2.81	0.275	141	287
42	2.95	0.289	143	290
44	3.09	0.302	144	292
46	3.23	0.316	145	294
48	3.37	0.330	147	296
50	3.51	0.344	148	298
52	3.66	0.358	149	300
54	3.80	0.372	150	303
56	3.94	0.386	151	304
58	4.08	0.400	152	305
60	4.22	0.414	153	307
62	4.36	0.427	154	309
64	4.50	0.441	155	311
66	4.64	0.454	156	312
68	4.78	0.469	157	314
70	4.92	0.482	158	316
75	5.27	0.517	160	320
80	5.62	0.551	162	324
85	5.98	0.586	164	327
90	6.33	0.621	166	330
95	6.68	0.655	168	334
100	7.03	0.689	170	337
105	7.38	0.724	172	340
110	7.73	0.758	173	344
115	8.09	0.793	175	347
120	8.49	0.832	177	350
125	8.70	0.852	178	352

ATOMIC SYMBOLS AND ATOMIC WEIGHTS

Although the term *relative atomic mass* is today more strictly correct than *atomic weight*, the older name has been retained here as it is still in wide current use.

Originally atomic weights were considered as the mass of the atom relative to the mass of an atom of hydrogen. Today the values are taken relative to the isotope, carbon–12. In the case of certain radioactive elements no value of atomic weight may be quoted accurately without knowledge of origin and the value given in parentheses is for the nuclide of the element which has the greatest known half life.

Element	Symbol	Atomic weight
Hydrogen	H	1.008
Helium	He	4.003
Lithium	Li	6.941
Beryllium	Be	9.012
Boron	B	10.81
Carbon	C	12.01
Nitrogen	N	14.01
Oxygen	O	16.00
Fluorine	F	19.00
Neon	Ne	20.18
Sodium	Na	23.00
Magnesium	Mg	24.31
Aluminium	Al	26.98
Silicon	Si	28.09
Phosphorus	P	30.97
Sulphur	S	32.06

Element	Symbol	Atomic weight
Chlorine	Cl	35.45
Argon	Ar	39.95
Potassium	K	39.10
Calcium	Ca	40.08
Scandium	Sc	44.96
Titanium	Ti	47.90
Vanadium	V	50.94
Chromium	Cr	52.00
Manganese	Mn	54.94
Iron	Fe	55.85
Cobalt	Co	58.93
Nickel	Ni	58.70
Copper	Cu	63.55
Zinc	Zn	65.38
Gallium	Ga	69.72
Germanium	Ge	72.59
Arsenic	As	74.92
Selenium	Se	78.96
Bromine	Br	79.90
Krypton	Kr	83.80
Rubidium	Rb	85.47
Strontium	Sr	87.62
Yttrium	Y	88.91
Zirconium	Zr	91.22
Niobium	Nb	92.91
Molybdenum	Mo	95.94
Technetium	Tc	(97)
Ruthenium	Ru	101.07
Rhodium	Rh	102.91
Palladium	Pd	106.4
Silver	Ag	107.87
Cadmium	Cd	112.40
Indium	In	114.82
Tin	Sn	118.69
Antimony	Sb	121.75
Tellurium	Te	127.60
Iodine	I	126.90
Xenon	Xe	131.30
Caesium	Cs	132.91
Barium	Ba	137.34
Lanthanum	La	138.91
Cerium	Ce	140.12
Praseodymium	Pr	140.91
Neodymium	Nd	144.24

Element	Symbol	Atomic weight
Promethium	Pm	(145)
Samarium	Sm	150.4
Europium	Eu	151.96
Gadolinium	Gd	157.25
Terbium	Tb	158.93
Dysprosium	Dy	162.50
Holmium	Ho	164.93
Erbium	Er	167.26
Thulium	Tm	168.93
Ytterbium	Yb	173.04
Lutetium	Lu	174.97
Hafnium	Hf	178.49
Tantalum	Ta	180.95
Tungsten	W	183.85
Rhenium	Re	186.21
Osmium	Os	190.2
Iridium	Ir	192.22
Platinum	Pt	195.09
Gold	Au	196.97
Mercury	Hg	200.59
Thallium	Tl	204.37
Lead	Pb	207.2
Bismuth	Bi	208.98
Polonium	Po	(209)
Astatine	At	(210)
Radon	Rn	(222)
Francium	Fr	(223)
Radium	Ra	226.03
Actinium	Ac	(227)
Thorium	Th	232.04
Protoactinium	Pa	231.04
Uranium	U	238.029
Neptunium	Np	237.05
Plutonium	Pu	(244)
Americium	Am	(243)
Curium	Cm	(247)
Berkelium	Bk	(247)
Californium	Cf	(251)
Einsteinium	Es	(254)
Fermium	Fm	(257)
Mendelevium	Md	(258)
Nobelium	No	(255)
Lawrencium	Lr	(260)

COMMON NAMES AND ABBREVIATIONS FOR PLASTICS

Many plastics are often known by abbreviations or by common names rather than by a true chemical name. This is usually acceptable as long as there is no ambiguity, as for example, the common use in Britain of the word polythene for the polymer of ethylene. In many cases however ambiguities do exist. For example, use of the word *urea* could mean either the chemical *urea*, a *urea-formaldehyde resin* or a *moulding material based on a urea-formaldehyde resin*. Similarly the word *acrylic*, may simply be an adjective used to describe a class of rubber, or fibre or plastic material or polymer in general. Many hundreds of acrylic polymers are known. However in the plastics industry the term acrylic is often taken to mean polymethyl methacrylate.

This ambiguity may be reduced by the use of standardized names and abbreviations. A number of such standards now exist including the following:

ISO 1043 – 1978. Plastics – Symbols.

BS 3502. Common Names and Abbreviations for Plastics and Rubbers.
Part 1. Principal commercial plastics (1978).
(The 1978 revision was carried out in accordance with ISO 1043 although the latter also deals with compounding ingredients.)

ASTM D 1600–83. Abbreviations of terms relating to plastics.

DIN 7728. Part 1 (1978). Symbols for terms relating to homopolymers, copolymers and polymer compounds.
Part 2 (1980). Symbols for reinforced plastics.

In the following list, drawn up by the author, of abbreviations in common use, those in **bold** type indicate that they are in the main schedule of BS 3502. In this list the names given for the materials are the *commonly used scientific names*. This situation is further complicated by the adoption of a nomenclature by the International Union of Pure and Applied Chemistry for systematic names and a yet further nomenclature by the Association for Science Education which is widely used in British schools but not in industry. Some examples of these are given in the second table.

Abbreviation	Material	Common name
ABS	acrylonitrile-butadiene-styrene polymer	ABS
ACS	acrylonitrile-styrene and chlorinated poly-ethylene	
AES	acrylonitrile-styrene and ethylene propylene rubber	
ASA	acrylonitrile-styrene and acrylic rubber	
CA	cellulose acetate	acetate
CAB	cellulose acetate-butyrate	CAB, butyrate
CAP	cellulose acetate-propionate	CAP
CN	cellulose nitrate	celluloid
CP	cellulose propionate	CP, propionate
CTA	cellulose triacetate	triacetate
CS	casein	casein
DMC	(usually polyester)	dough moulding compound
EP	epoxide resin	epoxy
ETFE	tetrafluoroethylene-ethylene copolymer	
EVAC	ethylene-vinyl acetate	EVA
EVOH, EVAL	ethylene-vinyl alcohol	
FEP	tetrafluorethylene-hexafluoropropylene copolymer	
FRP, FRTP	thermoplastic material reinforced, commonly with fibre	
GRP	glass fibre reinforced plastic based on a thermosetting resin	
HDPE	high density polyethylene	
HIPS	high impact polystyrene	
LDPE	low density polyethylene	
LLDPE	linear low density polyethylene	
MBS	methacrylate-butadiene styrene	
MDPE	medium density polyethylene	
MF	melamine-formaldehyde	melamine
PA	polyamide	nylon (some types)
PBTP, PBT, PTMT	polybutylene terephthalate	polyester
PC	polycarbonate	polycarbonate
PCTFE	polychlorotrifluoroethylene	
PE	polyethylene	polythene
PEBA	polyether block amide	

Abbreviation	Material	Common name
PEEK	polyether ether ketone	
PEG	polyethylene glycol	
PEK	polyether ketone	
PES	polyether sulphone	
PETP, PET	polyethylene terephthalate	polyester
PF	phenol-formaldehyde	phenolic
PMMA, PMM	polymethyl methacrylate	acrylic
POM	polyacetal, polyoxymethylene, polyformaldehyde	acetal
PP	polypropylene	propylene, polyprop
PPG	polypropylene glycol	
PPO	polyphenylene oxide	
PPO	polypropylene oxide	
PPS	polyphenylene sulphide	
PS	polystyrene	styrene
PS, PSU		polysulphone
PTFE	polytetrafluoroethylene	PTFE
PUR	polyurethane	polyurethane, urethane
PVA	polyvinyl acetate	
PVA	polyvinyl alcohol	
PVA	polyvinyl acetal	
PVAC	polyvinyl acetate	PVA
PVB	polyvinyl butyral	
PVC	polyvinyl chloride	PVC, vinyl
PVDC	polyvinylidene chloride	
PVDF	polyvinylidene fluoride	
PVF	polyvinyl fluoride	
PVF	polyvinyl formal	
PVP	polyvinyl pyrrolidone	
RF	resorcinol-formaldehyde	
SAN	styrene-acrylonitrile	SAN
SI	polysiloxane	silicone
SMC	(usually polyester)	sheeting moulding compound
TPS	toughened polystyrene	
UF	urea-formaldehyde	urea
UP	unsaturated polyester	polyester
UPVC	unplasticized PVC	
VLDPE	very low density polyethylene	
XPS	expanded polystyrene	

The Commission on Macromolecular Nomenclature of the International Union of Pure and Applied Chemistry have published a nomenclature for single strand organic polymers (*Pure and Applied Chemistry*, **48**, 375 (1976). In addition the Association for Science Education in the UK have made recommendations based on a more general IUPAC terminology and its suggestions have been widely used in British schools. Some examples of this nomenclature compared with normal usage are given below.

Normal usage	ASE	IUPAC
Polyethylene	poly(ethene)	poly(methylene)
Polypropylene	poly(propene)	poly(propylene)
Polystyrene	poly(phenyl ethene)	poly(1-phenyl ethylene)
Polyvinyl chloride	poly(chloroethene)	poly(1-chloroethylene)
Polymethyl methacrylate	poly(methyl 2-methyl propenoate)	poly[1-(methoxycarbonyl)-1-methyl ethylene]

In this Handbook the policy has been to use *normal usage scientific terms*.

DENSITY AND SPECIFIC GRAVITY OF POLYMERS

Density and specific gravity (sometimes known as relative density) are measures of the weight of a material per unit volume. Compared with metals, most plastics materials have a low density. This may be increased by the incorporation of denser fillers or reduced in the case of cellular, including foamed, materials. The data below, unless otherwise stated is for unfilled material.

Density is most conveniently expressed in g/cc, although the SI recommendations are for kg/m^3 (=0.001 g/cc). The density of water is 1.0 g/cc (at 4 °C). Since the specific gravity is the ratio density of material: density of water (and thus dimensionless), it is numerically equal to the density when the latter is expressed in g/cc.

Since plastics parts are made to a certain volume, whilst raw materials are bought according to their weight, it is often useful to determine the price per unit volume by multiplying the price per unit weight by the density (in the appropriate units).

The density of a polymer depends on the processing conditions, particularly so with crystalline polymers. It is also to be appreciated that in the melt the densities may be somewhat lower, again particularly so with crystalline polymers.

1 Mg/m^3 = 1 g/cc
= 62.43 lb/cu.ft = 0.03613 lb/cu.in

Solid polymer (Unfilled)	Density (g/cc) or specific gravity
Poly–4–methyl pentene–1	0.83
Polypropylene	0.90
Polyethylene	0.914–0.96
Polybut–1–ene	0.95
Ethylene–vinyl acetate	0.93–0.95
Ionomer	0.93
Ethylene–ethyl acrylate	0.93
ABS	1.01–1.07
Polystyrene	1.054
High-impact polystyrene	1.02–1.05
Styrene-acrylonitrile	1.06
PPO and PPO/PS blends (Noryl)	1.06
Nylon 12	1.02
Nylon 11	1.04
Nylon 610	1.09
Aromatic polyamide (Trogamid)	1.12
Nylon 6	1.13
Nylon 66	1.14
Nylon 46	1.18
Polymethyl methacrylate	1.17–1.19
Polyvinyl carbazole	1.19
Polycarbonate (of bisphenol A)	1.20
Polysulphone	1.24
Polyether ether ketone	1.265–1.32
Polyether–imide (Ultem 1000)	1.27
Polybutylene terephthalate	1.32
Polyether ketone	1.32
Polyphenylene sulphide	1.35
Polyethylene terephthalate	1.37
Polyethersulphone	1.37
Polyester-liquid crystal (Vectra)	1.40
Polyvinyl fluoride	1.38–1.57
Polyvinyl chloride	1.4
Acetal copolymer	1.410
Polyimide (Vespel)	1.42
Acetal homopolymer	1.425
Polyvinylidene fluoride	1.76
Polyvinylidene chloride	1.875
Polychlorotrifluoroethylene	2.1–2.3
Tetrafluoroethylene-hexafluoropropylene	2.16
Polytetrafluoroethylene (PTFE)	2.1–2.3

HEAT STABILITY OF POLYMERS

The data given in this section is concerned with chemical changes that occur on prolonged exposure of polymers to elevated temperatures. It is not concerned with softening which may occur on heating. The first table is largely based on data from *Properties of Polymers* by Van Krevelen and is an attempt to give estimates for a broad range of materials. In recent years there has been increasing interest in the Temperature Index of the Underwriters Laboratories. This is based on studies of the changes in values of certain properties over a range of elevated temperatures, from which the temperature at which a property loses half its value after 10 000 h is extrapolated. The resulting temperature index will depend on the property being measured and on the grade of material. Some typical data for temperature index are given in the second table.

Polymer	Ultimate end-use temperature 200 h/ °C	Ultimate end-use temperature 1000 days/ °C
Polyvinyl chloride	60–90	50
Polystyrene	60–90	60
Polyisoprene	60–90	60–80
Polymethacrylates	70–100	60–80
Polyolefins	70–100	60–90
Polyamides	100–150	80–100
Linear polyurethanes	130–180	70–110
Unsat. polyurethanes	130–220	80–110
Epoxy resins	140–250	80–130
X-linked polyurethanes	150–250	100–130
Polycarbonate	120	100–135
PET	140–200	100–135
X-linked ar. polyester	180–250	120–150
PPO	160–180	130–150
Polysulphone	160–180	130–150
Fluoroelastomers	200–260	130–170

Polymer	Ultimate end-use temperature 200 h/ °C	Ultimate end-use temperature 1000 days/ °C
Silicone elastomers	200–280	130–180
Polyesterimides	200–280	150–180
Polyamide imides	200–280	150–180
Silicone resins	200–230	150–200
Fluoroplastics	230–300	150–220
Aromatic polyamides	250–300	180–230
Polyimides	300–350	180–250
Polybenzimidazole	350–400	250–300

Continuous Use Temperature (Mechanical without impact)

Polymer	Temperature (°C)
ABS	60–80
Nylons	75
Polyacetal (homopolymer)	90
Polyacetal (copolymer)	90
PBT	105–140
PET	140 (glass-reinforced)
Polyarylate (Ardel D–100)	130
Polysulphones	160
Polyetherimide	170
Polyethersulphone	180
Polyphenylene sulphide	200 (glass-reinforced; mechanical with impact)
Ar. polyester (Ekkcel I–2000)	220
Liquid crystal polyester	220
Polyether ether ketone	240

The data given in this table has been obtained using procedures related to the UL Temperature Index method. The data given here does not however necessarily imply that the Underwriters Laboratories have awarded the temperature ratings quoted here either for the class of materials or for specific grades. Unless specified the data given is for the standard unreinforced polymers.

REFERENCE

Van Krevelen, D. W. (1976). *Properties of Polymers*. New York; London; Amsterdam: Elsevier.

FLAME RETARDANCY OF POLYMERS

―――――――――――――――――――――

―――――――――――――――――――――

LIMITING OXYGEN INDEX

The Limiting Oxygen Index is the minimum concentration of oxygen, expressed as a percentage, in a slowly rising stream of a mixture of oxygen and nitrogen which will support combustion. The percentage of oxygen in air is 20.9, so that polymers with a higher figure than this will not burn easily, if at all. Where only a single figure is given in the table it will apply to raw polymer, unless otherwise indicated; where there is a spread of figures this indicates that the result is dependent on the grade of material, which may be dependent on the polymer structure or the presence of additives such as glass fibre and flame retardants. The relevant standards are ASTM D–2863 and BS 2782 Method 141B.

Where plastics materials exhibit a measure of flame retardancy it is more common to classify them according to the Underwriters Laboratory UL 94 Vertical Burning Test (see next section).

Polymer	Limiting Oxygen Index (%)
Polyacetal	15
Polymethyl methacrylate	17
Polypropylene	17
Polyethylene	17
Polybutylene terephthalate	18
Polystyrene	18
Polyethylene terephthalate (unfilled)	21
Nylon 6	21–34
Nylon 66	21–30
Nylon 11	25–32
PPO	29–35
ABS	29–35

Polymer	Limiting Oxygen Index (%)
Polycarbonate of bisphenol A	26
Polysulphone	30
Polyethylene terephthalate (30% G.F)	31–33
Polyimide (Ciba-Geigy P13N)	32
Polyarylate (Solvay Arylef)	34
Polyether sulphone	34–38
Polyether ether ketone	35
Phenol–formaldehyde resin	35
Urea – formaldehyde resin	35
Polyvinyl chloride	23–43
Polyvinylidene fluoride	44
Polyamide-imides (Torlon)	42–50
Polyphenylene sulphide	44–53
Polyvinylidene chloride	60
Polytetrafluoroethylene	90

UNDERWRITERS LABORATORY UL 94 VERTICAL BURNING TEST

In this test a rectangular bar is held vertically and clamped at the top. The burning behaviour of the bar and its tendency to form burning drips when exposed to a methane or natural gas flame applied to the bottom of the specimen is noted. Methods of specimen preparation and details of the test method are laid down in Underwriters Laboratory Inc., Standard UL 94. Materials are rated 94 V–0, 94 V–1 or 94 V–2 according to the following characteristics:

94 V–0: Specimen burns for less than 10 s after application of test flame. The wad of surgical cotton beneath the specimen is not ignited. When the specimen stops burning the flame is replaced for a further 10 s; after removal of this flame, glowing combustion should die within 30 s. The total flaming combustion time for ten flame applications on five specimens should be less than 50 s.

94 V–1: Specimen burns for less than 30 s after either test flame application: cotton not ignited. Glowing combustion dies within 60 s

of second flame removal. Total flaming combustion time for ten flame applications on five specimens should be less than 250 s.

94 V–2: As for 94 V–1 except that there may be some flaming particles which burn briefly but which ignite the cotton.

The UL codings apply to a *specific grade* of material. The ratings may be dependent on the sample thickness conditions, being more stringent with thinner samples. Many of the so-called engineering thermoplastics now meet the V–0 rating and there has been an increasing tendency to quote minimum sample thickness for which the material will pass the V–0 test.

Mention should also be made of two other UL ratings; the 94HB (HB = horizontal burn) rating for which the sample passes a less severe burn test than the vertical test and the 94 5–V rating which is more severe than the V–0 rating and involves a test in which the sample is subjected to a 5 in. flame 5 times at 5 s intervals with each exposure lasting 5 s.

Whilst providing a guide these ratings are not intended to reflect hazards presented by the material under actual fire conditions.

Polymer		UL 94 Rating
Polycarbonate	Lexan 101	V–2
Nylon 66	Maranyl A100	V–2
Nylon 66 G.F.	Maranyl AD 447	V–0
Polysulphone	Udel P–1720	V–0
Polyethersulphone	Victrex 200P	V–0
Polyethylene terephthalate G.F.	Rynite 530 FR	V–0
PPO	Noryl HS 1000	V–0
	Noryl GFN–2	HB
Polyacetal	Delrin 500	HB
Polyphenylene sulphide G.F.	Ryton	V–0
Polyarylate	Ardel D–100	V–0
Polyether–imide	Ultem	V–0
Polyether ether ketone	Victrex PEEK	V–0
ABS standard grades		HB
ABS/polycarbonate alloy	Cycovin KMP	V–0
Urea formaldehyde mouldings		V–0

GLASS TRANSITION TEMPERATURE AND MELTING POINT

Polymer molecules are effectively frozen below the glass transition temperature (commonly known by the abbreviation T_g) and the material in mass is rigid. If the polymer is an *amorphous* one the material begins to become rubbery above the T_g and, providing the molecular weight is not too great further raising of the temperature (typically to about $T_g + 60°C$ will cause the material to melt. The T_g thus normally determines the upper service temperature limit of a plastics material in respect of retaining form stability. For such amorphous plastics, values for softening points are similar to the values for the T_g. Processing temperatures are usually some 50–100 deg C above the T_g.

Crystalline polymers retain their crystallinity above the T_g but melt over a range of temperature, the last traces of melting occurring at temperature about 50% higher than the T_g as expressed in degrees kelvin. This temperature is known as the crystalline melting point and usually referred to by the abbreviation T_m. The rigidity of the polymer between T_g and T_m will depend on the level of crystallinity. For unfilled crystalline polymers, common softening point tests give values intermediate to T_g and T_m, but glass-filled grades give softening points close to the T_m. Hence the T_m gives an indication of the maximum service temperature of a glass-filled polymer in terms of form stability. Melt processing is usually carried out some 30–70 °C above the T_m.

Three further points should be made.

1 The T_g depends to some extent on the manner by which it is measured so that the data given below can only be an indication.

2 For many polymers there may be additional transitions below the T_g at which limited molecular motion can occur and which may affect some properties (particularly electrical properties).

3 For some crystalline polymers the value of the T_g is in dispute and where this is the case the author has quoted the figure he believes to be most likely and marked the value with an asterisk *. Those further interested should see other publications by the author (Brydson, 1972: Brydson, 1989).

T_g and T_m Data for Some Common Polymers

Polymer	T_g (°C)	T_m (°C)
Polydimethyl siloxane	−123	−85 to −65
Natural rubber	−73	+25
Polyethylene	−20*	+100 to +126
Polypropylene	+5	+150
Polybut–1–ene	−20	+120
Poly–4–methyl pentene–1 (TPX)	+55	+245
Polytetrafluoroethylene (PTFE)	+115*	+327
Polychlorotrifluoroethylene (PCTFE)	+52	+221
Polyacetal (homopolymer)	−13*	+160
Polyvinyl chloride (UPVC)	+80	
Polystyrene	+90 to +100	
Polymethyl methacrylate	+99	
Polyethylene terephthalate (PET)	+67	+256
Polybutylene terephthalate (PBT)	+22 to +43	+224
Polyphenylene sulphide	+85	+285
Polycarbonate of bisphenol A	+149	+225
Polysulphone of bisphenol A	+190	
Polyether ether ketone	+144	+335
Polyether ketone	+154	+367
Polyarylate (Ardel)	+173 to +194	
Polyetherimide (Ultem)	+215	
Nylon 46	?	+295
Nylon 66	+60	+264
Nylon 6	+50	+215
Nylon 11	?	+185
Nylon 12	?	+175
Poly–m–xylylene adipamide	+85 to +100	+235

Brydson J.A. (Ed: A. D. Jenkins). (1972). *Polymer Science: A Materials Science Handbook*. (Chap. 3). Amsterdam; London: North Holland.
Brydson J. A. (1989). *Plastics Materials*. 5th ed. London: Butterworth.

SCALES OF HARDNESS

The hardness of a material is a complex property. In mineralogy it is usually considered in terms of scratch resistance, whereas with polymers the main concern is with resistance to indentation under load. Such indentation may be permanent or temporary, disappearing on removal of the load. Most standard tests do not normally distinguish between these two variants.

MOHS' SCALE OF HARDNESS

This is an arbitrary scale with irregular intervals in which ten materials are tabulated in ascending order with each material capable of being scratched by the mineral with the next higher hardness number.

Mineral	Mohs' hardness	Common equivalent
Talc	1	
Gypsum	2	
Calcite	3	Fingernail
Fluorite	4	
Apatite	5	Teeth, copper coin
Orthoclase	6	Window glass
Quartz	7	Penknife
Topaz and beryl	8	Hard file
Corundum	9	
(Emerald, sapphire, ruby)		
Diamond	10	

EXTENDED MOHS' SCALE

This scale has five additional points.

Mineral	Extended Mohs' hardness
Talc	1
Gypsum	2
Calcite	3
Fluorite	4
Apatite	5
Orthoclase	6
Vitreous silica	7
Quartz	8
Topaz	9
Garnet	10
Fused zirconia	11
Fused alumina	12
Silicon carbide	13
Boron carbide	14
Diamond	15

Comparative hardness scales for polymers
(Based on Van Krevelen, *Properties of Polymers*)

Pencil	Mohs'	Brinell	Rockwell		Shore			Types of product
			M	α (≈R)	D	C	A (≈IRHD*)	
7H	2	25	100					Hard plastics
		16	80					
2H		12	70	100	90			
		10	65	97	86			
HB		9	63	96	83			Moderately
		8	60	93	80			hard
B		7	57	90	77			plastics
		6	54	88	74			
2B	1	5	50	85	70			
		4	45		65	95		
		3	40		60	93	98	Soft
		2	32		55	89	96	plastics
		1.5	28		50	80	94	
		1	23		42	70	90	
		0.8	20		38	65	88	Rubbers
		0.6	17		35	57	85	
		0.5	15		30	50	80	
					25	43	75	
					20	36	70	
					15	27	60	
					12	21	50	
					10	18	40	
					8	15	30	
					6.5	11	20	
					4	8	10	

* IRHD = International rubber hardness degrees

SOLUBILITY PARAMETERS

It is frequently desired to have some preliminary information on solvents for a particular polymer and, as a corollary, what polymers will resist particular liquids. A very crude rule of thumb is that 'like dissolves like'. Rather more specific is the use of solubility parameters which are a measure of the forces holding molecules together. It is commonly observed that in the case of amorphous polymers and liquids the polymers will dissolve in solvents of similar solubility parameters.

This relationship does not necessarily hold for crystalline polymers which may be divided into two classes:

1 Crystalline polymers that are not capable of *specific interaction* with other chemicals including solvents. These include PE, PP and PTFE. None of these polymers dissolve at temperatures well below the crystalline melting points.

2 Crystalline polymers capable of specific interaction. In these cases it may be possible to find a solvent. Examples are phenol and glacial acetic acid which dissolve nylon 66. Polycarbonates and PVC are slightly crystalline and are more effectively dissolved by solvents capable of specific interaction.

USE OF THE TABLES

In the case of amorphous polymers a polymer will be expected to dissolve in a solvent with a value for the solubility parameter in the range ± 2 MPa$^{1/2}$ of the solubility parameter of the polymer. For example polystyrene ($\delta = 18.7$ MPa$^{1/2}$) will dissolve in such solvents as benzene (18.7), toluene (18.2), chloroform (18.7), cyclohexanone (20.2). (All figures in SI units.)

The symbol δ is very widely used to denote the solubility parameter. Much of the data quoted in the literature is in terms of (cal cm^{-3})$^{1/2}$ or MPa$^{1/2}$.

Crystalline polymers marked with a * are not capable of specific interaction and do not dissolve at room temperature, but may dissolve in solvents of similar solubility parameter when the crystalline melting point is approached.

Crystalline polymers marked with a † are capable of specific interaction with some solvents and where these solvents have similar values of δ solution may occur. In most of these cases it is necessary that one of the components (i.e. polymer and solvent) be what is known as a proton donor and the other a proton acceptor. Examples of proton donors are PVC, dichloromethane, formic acid, cresol and dimethyl formamide, while examples of proton acceptors are the nylons, polyethylene terephthalate, the polycarbonates and cyclohexanone.

Solubility Parameters of Polymers

Polymer	Solubility parameter (δ)	
	(cal cm^{-3})$^{1/2}$	(MJm^{-3})$^{1/2}$
Polytetrafluoroethylene *	6.2	12.6
Polychlorotrifluoroethylene	7.2	14.7
Polydimethyl siloxane	7.3	14.9
Ethylene–propylene rubber	7.9	16.1
Polyisobutylene	7.9	16.1
Polyethylene *	8.0	16.3
Polypropylene *	8.0	16.3
Polyisoprene (natural rubber)	8.1	16.5
Polybutadiene	8.4	17.1
Styrene-butadiene rubber	8.4	17.1
Poly–t–butyl methacrylate	8.3	16.9
Poly–n–hexyl methacrylate	8.6	17.6
Poly–n–butyl methacrylate	8.7	17.8
Polyethyl methacrylate	9.0	18.3
Polymethylphenyl siloxane	9.0	18.3
Polysulphide rubber	9.0–9.4	18.3–19.2
Polystyrene	9.2	18.7
Polychloroprene rubber	9.2–9.4	18.7–19.2
Polymethyl methacrylate	9.2	18.7
Polyvinyl chloride †	9.5	19.4

bisphenol A polycarbonate †	9.5	19.4
Polyvinylidene chloride *	9.8–12.2	20.0–25.0
Ethyl cellulose	8.5–10.3	17.3–21.0
Cellulose dinitrate	10.55	21.6
Polyethylene terephthalate †	10.7	21.8
Acetal resins †	11.1	22.6
Cellulose diacetate	11.35	23.2
Nylon 66 †	13.6	27.8
Polymethyl–α–cyanoacrylate	14.1	28.7
Polyacrylonitrile	14.1	28.7

Because of the difficulties in their measurement, the published figures for polymer solubility parameters range ±3% on either side of the average figure quoted.

Solubility Parameters of Common Solvents

Solvent	(cal/cm^3)$^{1/2}$	MPa$^{1/2}$
neo-pentane	6.3	12.8
Isobutylene	6.7	13.7
n-Hexane	7.3	14.9
Diethyl ether	7.4	15.1
n-Octane	7.6	15.5
Methyl cyclohexane	7.8	15.9
Ethyl isobutyrate	7.9	16.1
Di-isopropyl ketone	8.0	16.3
Methyl amyl acetate	8.0	16.3
Turpentine	8.1	16.5
Cyclohexane	8.2	16.7
2,2–Dichloropropane	8.2	16.7
sec-Amyl acetate	8.3	16.9
Dipentene	8.5	17.3
Amyl acetate	8.5	17.3
Methyl n-butyl ketone	8.6	17.6
Pine oil	8.6	17.6
Carbon tetrachloride	8.6	17.6
Methyl n-propyl ketone	8.7	17.8
Piperidine	8.7	17.8
Xylene	8.8	18.0
Dimethyl ether	8.8	18.0
Toluene	8.9	18.2
Butyl cellosolve	8.9	18.2
1,2–dichloropropane	9.0	18.3

Solvent	$(cal/cm^3)^{1/2}$	$MPa^{1/2}$
Mesityl oxide	9.0	18.3
Isophorone	9.1	18.6
Ethyl acetate	9.1	18.6
Benzene	9.2	18.7
Diacetone alcohol	9.2	18.7
Chloroform	9.3	19.0
Trichloroethylene	9.3	19.0
Tetrachloroethylene	9.4	19.2
Tetralin	9.5	19.4
Carbitol	9.6	19.6
Methyl chloride	9.7	19.8
Dichloromethane	9.7	19.8
Dichloroethane	9.8	20.0
Cyclohexanone	9.9	20.2
Cellosolve	9.9	20.2
Dioxane	9.9	20.2
Carbon disulphide	10.0	20.4
Acetone	10.0	20.4
n-Octanol	10.3	21.0
Butyronitrile	10.5	21.4
n-Hexanol	10.7	21.8
sec-Butanol	10.8	22.0
Pyridine	10.9	22.3
Nitroethane	11.1	22.6
n-Butanol	11.4	23.2
Cyclohexanol	11.4	23.2
Isopropanol	11.5	23.4
n-Propanol	11.9	24.2
Dimethyl formamide	12.1	24.7
Hydrogen cyanide	12.1	24.7
Acetic acid	12.6	25.7
Ethanol	12.7	26.0
Cresol	13.3	27.1
Formic acid	13.5	27.6
Methanol	14.5	29.6
Phenol	14.5	29.6
Glycerol	16.5	33.6
Water	23.4	47.7

ESTIMATION OF SOLUBILITY PARAMETERS USING SMALL'S METHOD

A useful method for estimating solubility parameters from structural formulae was devised by P. A. J. Small (*J. Appl. Chem*, **3**, 71 (1953)). Small compiled a list of *molar attraction constants (G)* for parts of a molecule. For a given molecule the parts are added up (ΣG), multiplied by the specific gravity (ρ) and divided by the molecular weight (M), i.e.

$$\delta = \rho \Sigma G/M$$

Small's values for the molar attraction constants yield values of δ in the older units of $(cal/cm^3)^{1/2}$.

Using the data in the table below the value of the solubility parameter for polymethyl methacrylate which has the structure

$$-CH_2-\underset{\underset{COO\,CH_3}{|}}{\overset{\overset{CH_3}{|}}{C}}-$$

will be as follows:

Now M (for the repeating unit) = 100, $\rho = 1.18$

2 CH$_3$ at 214	= 428
1 CH$_2$ at 133	= 133
1 COO at 310	= 310
1 >C< at −93	= −93
ΣG	= 778

$\therefore \delta = 1.18 \times 778/100 = 9.2\ (cal/cm^3)^{1/2}$

$= 18.7\ MPa^{1/2}$

Molar attraction constants at 25 °C

Group	G	
—CH_3	214	
—CH_2—	133	
—CH⟨	28	
⟩C⟨	−93	
CH_2=	190	
—CH=	111	
⟩C=	19	
CH≡C—	285	
—C≡C—	222	
Phenyl	735	
Phenylene (o,m,p,)	658	
Naphthyl	1146	
Ring (5-membered)	105–115	
Ring (6-membered)	95–105	
Conjugation	20–30	
H	80–100	
O (ether)	70	
CO (ketones)	275	
COO (esters)	310	
CN	410	
Cl (single)	270	
Cl twinned as in ⟩CCl_2	260	
Cl triple as in —CCl_3	250	
Br single	340	
I single	425	
CF_2	150	In fluorocarbons only
CF_3	274	In fluorocarbons only
S sulphides	225	
SH thiols	315	
ONO_2 nitrates	≈440	
NO_2 aliphatic	≈440	
PO_4 organic	≈500	
Si in silicones*	≈38	

* Estimated by H. Burrell *Interchem Rev.* **14** 3 (1955)

HEATING AND COOLING REQUIREMENTS OF THERMOPLASTIC MELTS

When processing thermoplastics materials by a melt process such as injection moulding the amount of heat required to bring the material up to processing temperature will depend on:

1 Amount of material. Since a moulding is made to a particular volume it is probably more useful when comparing materials to compare heat requirements per unit volume.
2 The specific heat of the material (over the range between ambient and melt temperatures).
3 Any latent heat associated with melting of crystal structures.
4 The temperature rise.

In principle, the heat required to bring the material up to its processing temperature may be calculated in the case of amorphous polymers by multiplying the mass of the material (W), by the specific heat (s) and by the difference between the required melt temperature and ambient temperature (ΔT). In the case of crystalline polymers it is also necessary to add the product of mass times latent heat of melting of crystalline structures (L). Thus if the density of the material is D, then the enthalpy or heat required (E), to raise volume V, to its processing temperature will be given by

$$E = (Ws\Delta T + WL)/D.$$

Specific heat varies with temperature and it is often simplest to measure the total heat (enthalpy) requirements per unit mass directly by such techniques as differential scanning calorimetry. If desired an estimate of the specific heat may be made by dividing the enthalpy change between any two temperatures by the difference between the two temperatures ΔT. If the range spreads over the melting point in the case of a crystalline polymer the specific heat estimate in this case will be inflated by the value $L/\Delta T$.

In the following table, which is designed primarily for injection moulding, the melt and mould temperatures quoted are taken from a paper by Whelan and Goff.* From data obtained by differential scanning calorimetry methods, these authors then measured the amount of heat required to cool a unit mass of polymer from the melt temperature to the mould temperature, and used this as a measure of the heat required to be removed from a unit mass of a moulding before the moulding was extracted. In practice the mould is opened before the polymer reaches the mould temperature, but the data does provide useful information. Since a moulding is made by volume, this author has felt it more useful to compare the heat required to be removed per unit volume, and this has been obtained by using specific gravity data.

In their paper, Whelan and Goff estimated the specific heat averaged between the melt and mould temperatures by the method described above. It has been assumed by this author that this value is valid between the melt temperature and 20 °C, and has then been used to estimate the amount of heat required (both per unit volume and unit mass) to raise the polymer from 20 °C to the melt temperature. While this will be in some error, it once again provides some useful relative information.

The considerable difference between polymers should be noted, particularly that the polymers with the highest processing temperatures do not necessarily require the greatest amount of heat to raise to processing temperature, and in addition often require much less heat to be removed before a moulding can be extracted.

Heat required (enthalpy required) to raise polymers to their processing temperatures from an ambient temperature of 20 °C, and the heat required to be removed in cooling a polymer from the melt to mould temperature

Polymer	Melt tem-perature	Mould tem-perature	SG	Specific heat	Heat required to melt		Heat required to cool	
	°C	°C		J/kgK	J/g	J/cc	J/g	J/cc
FEP	350	220	2.2	1600	528	240	240	109
Polyethersulphone	360	150	1.37	1150	391	285	242	177
Polyether ether ketone	370	165	1.3	1340	469	361	275	212
Polyethylene terephthalate (cryst)	275	135	1.38	2180	556	403	305	221
Polystyrene	200	20	1.05	1720	310	295	310	295
Polyacetal	205	90	1.41	3000	555	394	345	245
Polycarbonate	300	90	1.2	1750	490	408	368	307
ABS	240	60	1.04	2050	451	434	369	355
Polymethyl methacrylate	260	60	1.18	1900	456	386	380	322
Polyphenylene sulphide	320	135	1.4	2080	624	446	385	275
PPO (Noryl-type)	280	80	1.06	2120	551	520	434	409
Polysulphone	360	100	1.24	1675	570	459	436	351
Polyethylene terephthalate (amorphous)	265	20	1.34	1970	483	360	483	360
Nylon 11/12	260	60	1.03	2440	586	568	488	474
LDPE	200	20	0.92	2780	500	543	500	543
Nylon 6	250	80	1.13	3060	703	623	520	460
Nylon 66	280	80	1.14	3075	800	701	615	539
Polypropylene	260	20	0.91	2790	670	736	670	736
HDPE	260	20	0.96	3375	810	843	810	843

* Whelan, A., Goff, J. P. Paper presented to the PRI Mouldmaking 1986 Conference at Solihull, England, January 1986.

THERMAL DATA RELEVANT TO DETERMINATION OF COOLING TIMES

The use of this data, particularly for assessing cooling times, is illustrated in Appendix 1 on specimen calculations. The freeze-off constant (B) is a crude factor relating the section thickness of a moulding (x) with the time (t) in seconds required for cooling in the mould before mould opening. The equation used is

$$t = Bx^2$$

The theory for the use of this equation is to be found in the booklet for Unit 7 of the Open University Course, Number PT614, Polymer Engineering.

Polymer	Process temperature range °C	Demoulding temperature °C	Thermal conductivity (@ Process temperatures) W m^{-1} K^{-1}	Thermal diffusivity 10^{-7} m^2 s^{-1}	Freeze-off constant Ms m^{-2}
LDPE	150–290	60	0.17	1.1	5.7–33
HDPE	190–290	70	0.2	1.1	2.8–6.9
PVC (plasticized)	150–190	30	0.15	0.7	15.1–25.8
PVC (unplasticized)	160–210	110	0.16	1.2	1.8–3.7
PP	200–290	105	0.15	1.1	4.2–11
POM	190–240	105	0.2	0.9	4.5–9.2
PS	200–260	100	0.16	0.8	7–14
ABS	200–260	110	0.16	0.9	4.3–8.5
PMMA	230–280	130	0.17	0.9	3.5–5.7
PPO	250–300	150	0.17	1.0	2.5–3.9
PC	290–320	170	0.18	1.0	2.5–3.2
Nylon 66	270–310	210	0.21	1.3	0.9–1.3
PES	320–400	220	0.18	1.1	1.6–2.8
PEEK	370–400	280	–	1.2	1.1–1.4

MOULDING SHRINKAGE DATA

Thermoplastic injection mouldings are slightly smaller than the cavities in which they are made. Such shrinkage may be expressed volumetrically or more commonly as a shrinkage in one direction (linear shrinkage). The shrinkage that occurs is the difference between two, opposing, effects:

1 Natural shrinkage on cooling determined by the coefficient of thermal expansion;
2 High moulding pressures which would actually cause a moulding to expand if the mould was opened when the polymer was still at its melt temperature.

The following general points should be noted:

1 Most amorphous thermoplastics have low values for linear moulding shrinkage of the order of 0.005 cm/cm. In the case of crystalline thermoplastics the process of crystallization on cooling leads to increased packing of the molecules and shrinkage is higher.
2 Shrinkage with filled materials may be less, depending on the coefficients of expansions of the additives used. Glass fibre filled compounds usually have a lower shrinkage.
3 Because of molecular orientation shrinkage in the flow direction may be different than in the directions transverse to the flow.
4 If the glass transition temperature is a little below normal ambient temperature, *aftershrinkage* may occur as the polymer mass takes time to equilibrate to size. In some cases this process may continue for up to 2 years. Stabilization of the mouldings can be effected by heating in a non-oxidizing medium for a short white at some temperature at least 50 °C above the glass transition point.

Polymer	Linear shrinkage (cm/cm; in/in)
ABS	0.004–0.008
Cellulosics (CA, CAB, CP)	0.003–0.008
Fluorinated ethylene propylene (FEP)	0.030–0.060
Nylon 6 (unfilled)	0.010–0.015
Nylon 66 (unfilled)	0.010–0.020
Nylons 11 and 12	0.005–0.020
Polyacetals	0.020–0.035
PBT (unfilled)	0.015–0.020
PBT (30% glass fibre)	0.003–0.008
PET (amorphous)	0.004 (flow direction)
	0.002 (transverse)
(crystalline)	0.018–0.021
(crystalline; 36% glass fibre)	0.002 (flow direction)
	0.018 (transverse)
Polycarbonate (unfilled)	0.006–0.008
Polycarbonate (30% glass fibre)	0.003–0.005
Polyether ether ketone	0.012–0.020
Polyethylene	0.015–0.060
Polymethyl methacrylate (acrylic)	0.004–0.007
Polyphenylene oxide (modified, e.g. Noryl)	0.005–0.007
Polyphenylene sulphide (g.f. filled)	0.002
Polypropylene	0.020
Polystyrene	0.004–0.005
Polysulphones and polyethersulphones	0.006–0.007
(40% glass fibre)	0.002
Polyvinyl chloride (unplasticized)	0.004
Polyvinylidene fluoride	0.023–0.030
(glass fibre filled)	0.020
Styrene-acrylonitrile (SAN)	0.004–0.005

SOME VALUES FOR THE FLOW PATH RATIO OF INJECTION MOULDING MATERIALS

The flow path ratio (or more accurately the flow path: wall thickness ratio) is basically a somewhat crude way of trying to assess the ease of flow of a polymer melt in an injection moulding machine. Because flow rates and heat transfer behaviour are not simply proportional to the wall thickness of the mould cavity, it might be expected that the flow path ratio would be of very limited use. However, many injection mouldings have similar cross-sections and indeed the writer has pointed out elsewhere (Brydson, J. A. (1989). *Plastics Materials*. 5th ed. London: Butterworth) that flow path ratio data does correlate surprisingly well with data obtained by more sophisticated techniques. The values do however depend on the melt temperature and the grade of polymer used (particularly the molecular weight), so that a spread of values is given.

Numerically the flow path ratio is the maximum length of flow found to be possible in an injection mould expressed in multiples of the wall thickness of the moulding (i.e. mould cavity cross-section).

Polymer	Flow path ratio
ABS	80–150
Acrylic [Poly(methyl methacrylate)]	100–150
Nylon 6	140–340
Nylon 66	180–350
Polyacetals	100–250
Poly(butylene terephthalate)	160–200
Polycarbonates	30–70
Polyether ether ketone	up to 200
Polyethylene (HDPE)	150–200
Polyethylene (LDPE)	200–300
Poly(ethylene terephthalate)	up to 350
Poly(phenylene sulphide)	150
Polypropylene	150–350
Polystyrene	150
Polystyrene (toughened)	130
Polysulphones	30–150
Poly(vinyl chloride) (plasticized)	up to 180
Poly(vinyl chloride) (unplasticized)	60
Styrene-acrylonitrile	140

ELECTRICAL CIRCUIT AND WIRING DIAGRAM SYMBOLS (FROM B.S.3939)

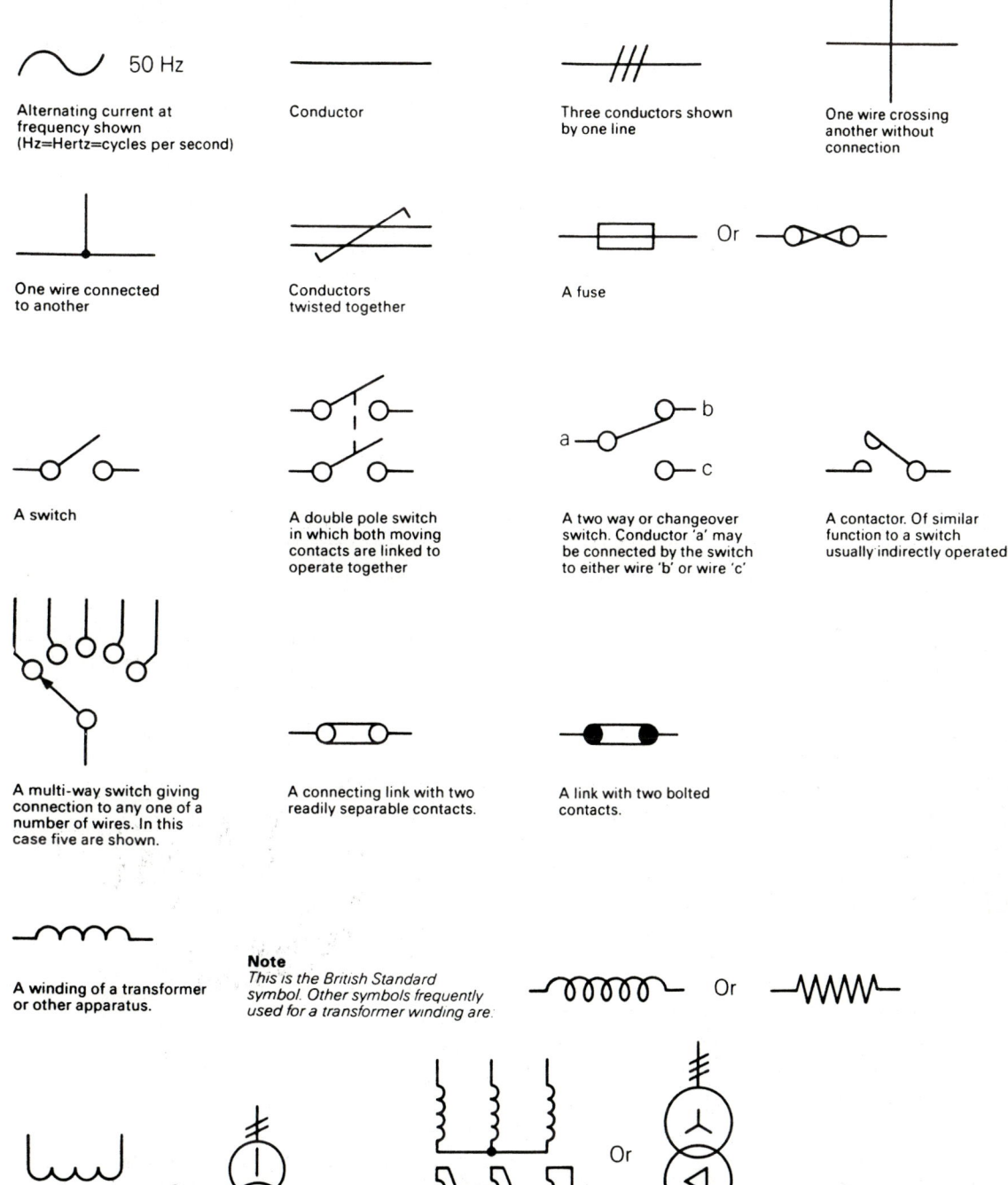

50 Hz

Alternating current at frequency shown (Hz=Hertz=cycles per second)

Conductor

Three conductors shown by one line

One wire crossing another without connection

One wire connected to another

Conductors twisted together

A fuse

Or

A switch

A double pole switch in which both moving contacts are linked to operate together

A two way or changeover switch. Conductor 'a' may be connected by the switch to either wire 'b' or wire 'c'

A contactor. Of similar function to a switch usually indirectly operated.

A multi-way switch giving connection to any one of a number of wires. In this case five are shown.

A connecting link with two readily separable contacts.

A link with two bolted contacts.

A winding of a transformer or other apparatus.

Note
This is the British Standard symbol. Other symbols frequently used for a transformer winding are:

Or

A single-phase transformer with separate windings.

Or

A three-phase transformer. (In this diagram the connections are star-delta).

HYDRAULIC SYMBOLS

Standard hydraulic symbols are to be found in BS 2917: 1977 and ISO 1219: 1976. Some of those more commonly found in plastics processing machinery drawings are given below.

Pump

Constant delivery volume

Variable delivery volume

Hydraulic motor
Constant motor

Adjustable motor

Hydraulic cylinder
Single operation

Double operation

Pressure valve
Valve, normally closed

Pressure limiting
valve fixed setting

Pressure reducing valve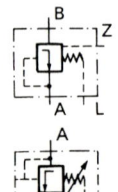

Sequence valve

Throttle valve
Throttle, constant

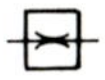

Throttle, adjustable

Flow control valve
Flow control valve fixed

Flow control valve variable

Directional valve
Check valve

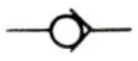

Control valve
4/2 Directional valve
operated by spool
return by spring

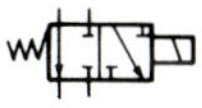

4/3 Directional valve
operated by spool
centered by spring

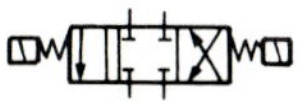

4/2 Directional valve
hydraulically operated

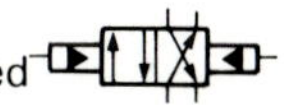

Sundries
Filter

Oil cooler

Cartridge

With internal
piloting

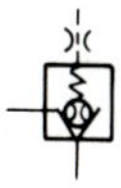

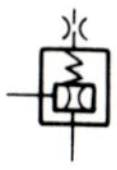

Cartridge valve Piston valve

With external
pilot control

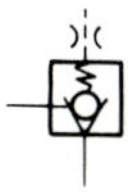

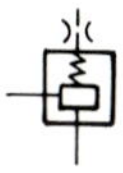

Supply line

Control line

Leakage line

Electrical line

Flexible connection

Conductor connection

Conductor crossover

FLOW FORMULAE

—————————————

—————————————

Most polymer melts approximate in their behaviour to power law fluids which in shear exhibit the relationship

$$\tau = K(\dot{\gamma})^n.$$

However this is only an approximation and for scale-up and other calculations it is often more useful to use flow curves based on shear stress against apparent shear rate which make no assumptions about the relationship between τ and $\dot{\gamma}$. Since the values of these variables at the wall of the flow channel are related to such measurable quantities as output, pressure drop between the ends of the channel and the dimensions of the channel, the data has wide applicability.

The formulae opposite give basic relationships for flow in a tube (capillary), a narrow slit and a narrow annulus.

Notes:

1 Q is the output rate, P is the pressure drop between ends of the channel, r is the tube raddius, L is the length of channel, w is the width of slit, H is the depth of slit and $\partial P/\partial Z$ is the rate of change of pressure along the channel and is equal to P/L.

2 It is assumed that $w \geqslant H$ and $(R_0 - R_1) \leqslant R_0$.

3 In practice n, n' and n'' are virtually identical. Strictly, n is the power law exponent in the power law formula, n' is the slope of the log-log plot of τ against $\dot{\gamma}$ for a capillary and n'' the corresponding figure for a slit.

4 Two definitions for apparent viscosity are widely used

either $\mu_a = \tau_w/\dot{\gamma}_w$
or $\quad\mu_a = \tau_w/\dot{\gamma}_{w,a}$

	Tube	Slit	Annulus
Shear stress at wall τ_w	$PR/2L$	$PH/2L$	$P(R_o - R_i)/2L$
Apparent shear rate at wall $\dot{\gamma}_{w,a}$	$4Q/\pi R^3$	$6Q/WH^2$	$6Q/\pi(R_o+R_i)(R_o-R_i)^2$
True wall shear rate $\dot{\gamma}_w$	$\dfrac{1}{\pi R^3}\left(3Q + P_1\dfrac{dQ}{dP}\right)$ $= \left(\dfrac{3n'+1}{4n'}\right)\dfrac{4Q}{\pi R^3}$	$\dfrac{2}{WH^2}\left(2Q + P\dfrac{dQ}{dP}\right) = \left(\dfrac{2n''+1}{3n''}\right)\left(\dfrac{6Q}{WH^2}\right)$	$2\dfrac{(2n''+1)}{n''}\cdot\dfrac{Q}{\pi(R_o+R_i)(R_o-R_1)^2}$
Average velocity vz	$\left(\dfrac{n+1}{3n+1}\right)v_o$	$\left(\dfrac{n+1}{2n+1}\right)v_o$	–
Maximum velocity (along flow centre line) v_o	$-\left(\dfrac{n}{n+1}\right)\left[\dfrac{1}{2k}\left(\dfrac{\partial P}{\partial Z}\right)\right]^{1/n} R^{(n+1)/n}$	$-\left(\dfrac{n}{n+1}\right)\left[\dfrac{1}{K}\left(\dfrac{\partial P}{\partial Z}\right)\right]^{1/n}\left(\dfrac{H}{2}\right)^{(n+1)/n}$	–

DATABASES

The increased use of computers has led to the proliferation of databases as well as information retrieval systems. Databases on material properties are available from raw material suppliers (e.g. ICI, Hoechst, BASF) and from research organizations (e.g. RAPRA). Other programs simulate aspects of mould processing such as mould filling (as in the Mouldflow system), mould cooling, extrusion and blow moulding. These are particularly useful in predicting how polymeric materials will behave in equipment of different geometries and under different conditions of operation. It is thus possible to predict the conditions required for optimum results at the 'drawing board' stage.

It should be stressed that simulated programs provide only a prediction – like a weather forecast, they may prove inaccurate either because the model used in the simulation is inadequate or because the data supplied is deficient. Thus, for example, while many simulations take into account the effects of temperature and shear rate on viscosity of a polymer melt, they may ignore the effects of such factors as compressibility of melts and elongational flow.

However, there have been many instances of simulation predictions that have been found to be accurate and have saved much time and expense in the design and commissioning of equipment such as moulds and dies. Furthermore, as the theoretical model becomes more refined and the supporting data more accurate, such simulations become increasingly reliable and effective.

APPENDIX

SPECIMEN CALCULATIONS FOR PROCESSORS

The plastics processor may be involved in all sorts of numerical calculations. This section deals with some of the most commonly encountered. For each calculation there is a brief theoretical explanation followed by a simple example.

It cannot be stressed too often that the properties of a polymer can rarely be quantified by a simple number, or relationships between properties by a simple formula. Thus calculations must be regarded as best estimates and wherever possible such estimates should be confirmed by direct experimental measurement.

In the case of two sets of calculations the reader is introduced to graphical methods of solution which can be particularly useful.

HYDRAULICS

Hydraulics may be defined as the practical application of hydrodynamics to engineering, the mathematical study of the motion, energy and pressure of liquids in motion. Because of its importance in polymer processing, the first few specimen calculations will also refer to some examples of hydrostatics where the fluid is at rest but exerting pressure.

SPECIMEN CALCULATION H1: TRANSMISSION OF PRESSURE IN A CLOSED SYSTEM

Theory

This problem invokes Pascal's Law of Fluid Pressures which states that pressure applied anywhere to an enclosed body of fluid is transmitted equally in all directions. This pressure acts at right angles to every portion of the surface of a container, the force per unit area being constant throughout.

Example

Two cylinders are connected by a pipe and are of the dimensions shown. They are loaded with water as indicated and pistons are mounted above the water in each cylinder. If a load of 2 kgf is applied to the small piston, what force may be exerted by the piston in the larger cylinder.

Since the area of the small piston is 4 cm² the *pressure exerted by the small piston on the water* is given by

Pressure on water = Load on piston/Area of
piston
= 2 kgf/4cm²
= 0.5 kgf/cm²

∴ Pressure on large piston = 0.5 kgf/cm²
∴ Force on large piston = 0.5 × 50 kgf = 25 kgf
∴ Force exerted by large piston = 25 kgf

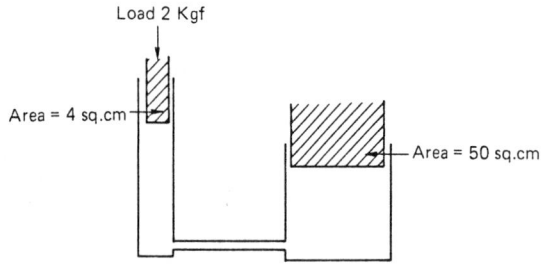

SPECIMEN CALCULATION H2: CALCULATION OF MOULDING PRESSURE IN A SIMPLE COMPRESSION MOULD

Theory

The basic system is as indicated in the diagram, the press being closed by fluid delivered by a pump under a *line pressure L*. The area of the hydraulic ram is A_r. It is assumed that when the press is closed, all of the force on the hydraulic ram is exerted on the material in the mould cavity, whose projected area (the area in the direction normal to the direction of ram movement) is A_m.

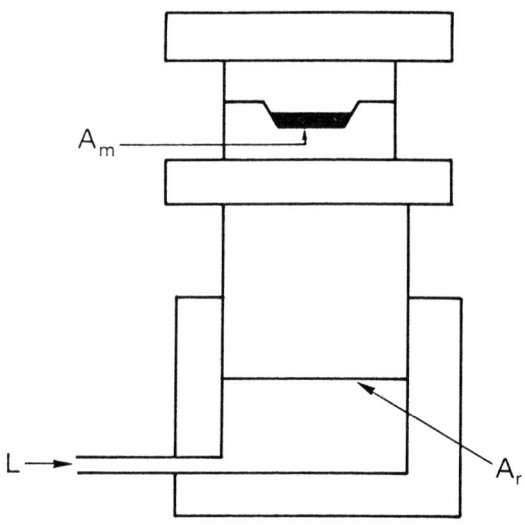

Force on hydraulic ram $= L \times A_r$
∴ Force on material in mould cavity $= L \times A_r$
Moulding pressure (M) $= L \times A_r/A_m$

Example

If the line pressure is 2240 p.s.i., the diameter of the hydraulic ram 12 inches and the projected area of the mould cavity 18 in², the moulding pressure will be given by

$$2240 \times 6 \times 6 \times \pi/18 = 14\,074.335 \text{ p.s.i.} = 6.28 \text{ t.s.i.},$$

where t represents a long ton of 2240 pounds.

(In compression moulding of thermosets which evolve volatiles during cure moulding, pressure is usually of the order of 1-2 t.s.i. In transfer moulding, the figure may be three times as high, whereas with polymers which do not evolve volatiles it is possible to work at very much lower moulding pressures.)

SPECIMEN CALCULATION H3: LINE PRESSURE, INJECTION PRESSURE AND LOCKING FORCE RELATIONSHIPS IN INJECTION MOULDING

Theory

The line pressure (L) is the pressure of the hydraulic oil acting on the hydraulic piston at the rear of the injection screw.

At the present time a great diversity of units are used in connection with pressure, such as pounds per square inch (p.s.i.), kilogrammes force per square centimetre (kgf cm^{-2}), kilopons per square centimetre (kp cm^{-2}), bars, meganewtons per square metre (MN m^{-2}) and megapascals (MPa).

The relationship between these is given for convenience here.

$$1 \text{ MPa} = 1 \text{ MN m}^{-2} = 1000 \text{ kN m}^{-2}.$$

The following relationships are approximate (see table on page 157 for accurate conversions)

$$1 \text{ MN m}^{-2} \simeq 10 \text{ kgf cm}^{-2} \simeq 10 \text{ kp cm}^{-2} = 10 \text{ bar}$$
$$= 145 \text{ p.s.i.}$$

Example   The following conditions exist in an injection moulding machine: screw diameter = 5 cm, piston diameter = 15 cm, line pressure = 14 MPa

$\therefore$ Injection pressure = $(7.5 \times 7.5 \times 14)/(2.5 \times 2.5)$
                                = 126 MPa (i.e. $\simeq$ 18 300 p.s.i.).

Let us assume for a moment that there is no pressure loss in the molten plastics material between the injection screw and the blind end of the mould cavity. In such case, Pascal's Law would apply and the force tending to open the mould would be given by $I \times M$, where I is the injection pressure and M the projected mould cavity area. If M were 150 in^2 then the mould opening force, assuming no pressure loss, would be

150 $\times$ 18 300 pounds = 2 745 000 pounds $\simeq$ 1200 tons.

In practice there will be considerable pressure drops in the melt and the procedures given in H4 should be used.

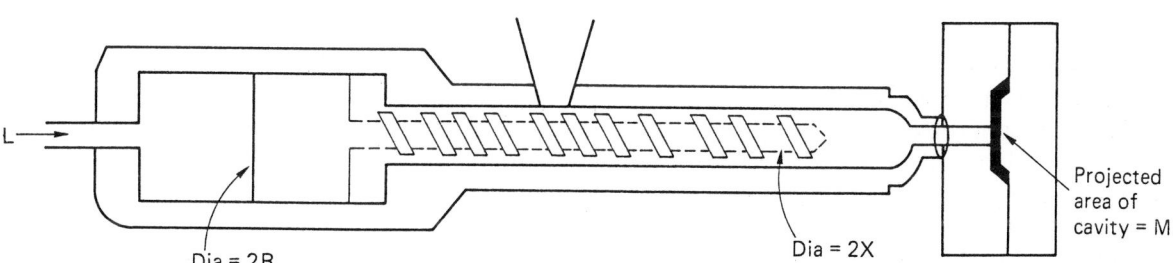

Let the diameter of the piston at the rear of the injection screw be $2R$

$\therefore$ Force on rear of screw = $\pi R^2 L$.

This force will be exerted by the front of the screw on the polymer melt. If the screw has a diameter of $2X$, i.e. cross-sectional area of πX^2,

$\therefore$ Pressure on plastics melt (the *injection pressure*) will be given by

$$R^2 L / X^2.$$

SPECIMEN CALCULATION H4: CALCULATION OF MOULD LOCKING FORCES

It is useful to distinguish between the mould opening force developed up to the moment that the mould cavity fills (i.e. at the last moment that the material is still flowing), and the mould opening force during the packing stage immediately after mould filling.

In the first instance, there will be pressure losses due to flow all along the line from the front of the injection ram, through the barrel, nozzle, sprue, runner, gate and mould cavity. These pressure losses will be due to shear flow and, where there is converging flow such as in the nozzle areas, tensile flow. Whilst it is possible to add all of the pressure losses at each stage to both of these types of flow, it is generally simplest to divide the total pressure drop (P_m) into two components, the delivery pressure which is the pressure drop between the front of the ram and the gate of the mould cavity (ΔP), and the pressure drop in the mould cavity at the instant the mould just fills (p_0), i.e.

$$P_m = \Delta P + P_0$$

ΔP may be determined experimentally by measuring the pressure generated in an air shot at the correct volumetric injection rate and adding the pressure drop generated through the runners determined in the next example. Hence for a known injection pressure (P_m), P_0 may be obtained simply by subtracting ΔP.

In one analysis (Barrie, 1970) it has been suggested that for a centre gated disc the mould opening force at the moment of mould fill will be given by

$$F = P_0 \, \pi R^2 \, (n/n+2),$$

where n is an experimentally derived factor not necessarily equated with the flow behaviour index. (A much more detailed analysis may be found in the Open University, PT 614, Polymer Engineering Booklet entitled Unit 7, Injection Moulding.)

During the packing stage there should, in theory, be complete transmission of pressure between ram and blind end of the mould cavity up until the moment the gate freezes. In practice there is some pressure loss and hence the mould opening force may be expressed by the equation.

$$F = PAG,$$

where P is the injection pressure, A is the projected area of the moulding and G is a correction factor (usually 0.8–0.9 but as low as 0.7 in extended runner systems).

Comparison of the two equations for mould opening force show that, at least in the case of a centre gated disc mould, the mould opening force, and hence locking force requirements will be much higher during the packing stage (Barrie, I. T. (1970). *Plastics and Polymers* **37**, 463.

SPECIMEN CALCULATION H5: FLOW THROUGH RUNNERS AND GATES

It was shown in Chapter 4 that the shear stress and apparent shear rate at the wall of a tube were given respectively by

$$\tau_w = PR/2L$$

and

$$\dot{\gamma}_w = 4Q/\pi R^3.$$

Let us consider an injection mould in which the cavity is fed by a full-round runner 70 mm in length and of radius 2.5 mm. For simplicity of calculation let us also assume that the gate is full-round and of radius 0.5 mm and length 1 mm. Let us also assume that we have a two-cavity mould and that during the injection stage the ram displaces 500 cm^3 s^{-1}.

To calculate the shear rate in the runner:
  Assuming there are separate runners from the sprue to the two cavities the flow rate up each runner Q will be

$$500/2 = 250 \text{ cm}^3 \text{ s}^{-1}$$

The shear rate at the wall will therefore be

$$4 \times 250/\pi \times 0.25^3 = 2.04 \times 10^4 \text{ s}^{-1}.$$

To calculate the shear rate at the wall of the gate:
  The flow rate will be the same, so that

$$\dot{\gamma}_w \text{ (gate)} = 4 \times 250/\pi \times 0.05^3 = 2.55 \times 10^6 \text{ s}^{-1}$$

To calculate the pressure drop (P) from one end of the runner to the other:
  We first read off the shear stress corresponding to the shear rate for the appropriate material from the flow curve. The figure illustrates some hypothetical material at the appropriate temperature. The value obtained is 0.7 MPa.

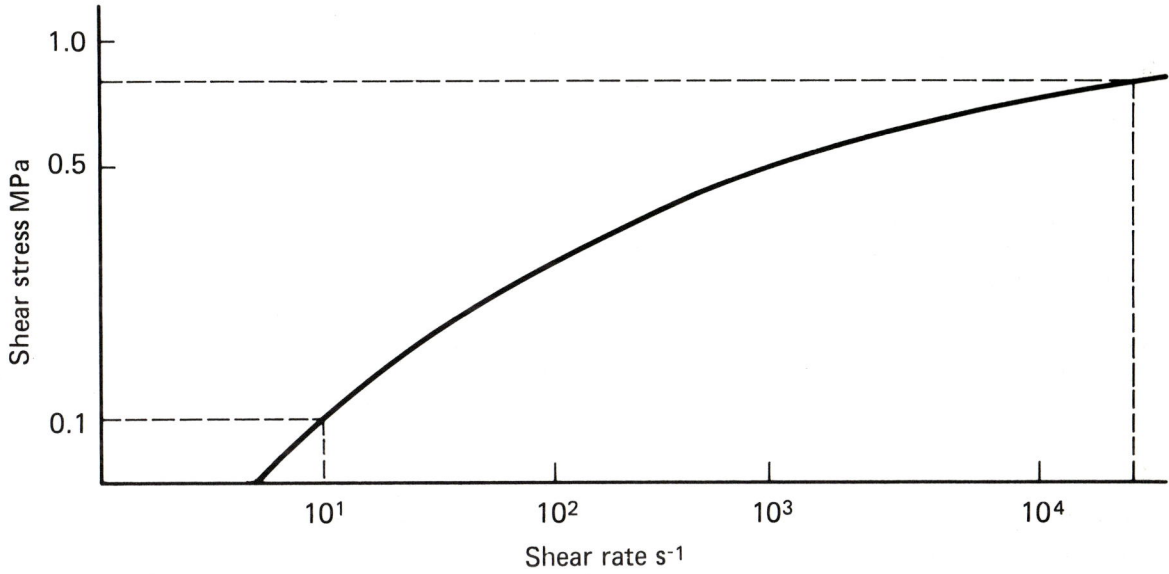

Now the shear stress at the wall $\tau = PR/2L$, hence by rearranging

$P = 2L\tau/R$
$= 2 \times 70 \times 0.7/2.5$
$= 39.3$ MPa
$= 39\ 300\ 000$ N/m²
or 5700 p.s.i.

To calculate the viscosity at this shear stress at the wall:

$$\text{Viscosity} = \tau/\dot{\gamma} = 700\ 000/20\ 400 = 34\ \text{N s m}^{-2}$$

To calculate the viscosity at a shear rate of 10 s⁻¹:

Read off the corresponding shear stress. This is 0.1 MPa. Hence

$$\text{viscosity} = \tau/\dot{\gamma} = 100\ 000/10 = 10\ 000\ \text{N s m}^{-2}$$

This illustrates the highly pseudoplastic nature of the melt.

HEAT TRANSFER

SPECIMEN CALCULATION: MOULD COOLING

It is often required to estimate cooling times in injection moulding. This is an exercise in conductive heat transfer the theory for which can be very complicated. In many cases, however, it is possible to use a simple graphical method based on theory, but which does not involve the use in any detailed knowledge of the theory.

The method involves the use of a cooling curve and graphs are given below both for a cylindrical shape and for a flat sheet. The calculation given will use only the curve for the flat sheet.

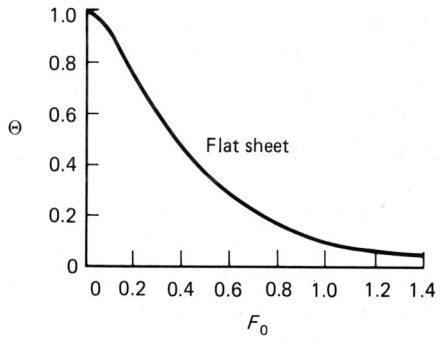

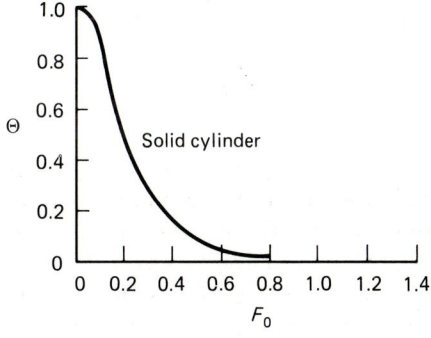

The cooling curve has as its axes two dimensionless parameters, the dimensionless temperature (θ) and the Fourier number F_0 which are given by

$$\theta = \frac{T - T_E}{T_0 - T_E}$$

where T_0 is the initial melt temperature, T_E is the temperature of the cooling environment and T is the temperature to which the polymer has cooled after time t and

$$F_0 = \frac{at}{x^2}$$

where a is known as the *thermal diffusivity* and is in turn given by

$$a = k/\rho c$$

where k is the thermal conductivity, ρ the polymer density and c the specific heat, t is the time in seconds and x the diffusion path length in the cooling process.

Example

A flat plaque of polystyrene is being moulded with a wall thickness of 0.4 cm. The initial melt temperature is 200 °C, the mould temperature is 60 °C and the moulding may be extracted when the centre of the moulding has reached its glass transition temperature of 85 °C. The thermal diffusivity of the material is $8.35 \times 10^{-8} \text{ m}^2 \text{ s}^{-1}$. In these circumstances θ may be estimated as $85 - 60/200 - 60 = 0.18$. From the flat sheet cooling curve this is seen to correspond to a Fourier number (F_0) of 0.78.

From the formula for the Fourier number

$$t = x^2 F_0 / a.$$

In this example the diffusion path length will be the distance from the centre of the moulding to the wall, i.e. half the thickness or 0.2 cm = 0.002 m.

Hence the required cooling time

$$t = 0.002^2 \times 0.78/8.35 \times 10^{-8}$$
$$= 38 \text{ s}.$$

It is perhaps worth emphasizing that halving the wall thickness will reduce the cooling time to a quarter of its previous value, i.e. a mould-ing of wall thickness 0.2 cm will require a cooling time of 9.5 s.

The cooling time may also be estimated from the data on page 181 of the data section using the equation

$$t = Bx^2.$$

The value for B of 7 M s m^{-2} is based on a demoulding temperature of 100 °C and a melt temperature of 200 °C. For consistency of units the value of x in metres will be 0.04. Hence

$$t = 7 \times 0.004 \times 0.004 \text{ M s}$$
$$= 112 \text{ s}.$$

Clearly there is a large difference in the times. It has been suggested (see Open University Course, PT614, Polymer Engineering, Unit 7, booklet) that the cooling curve method seems to underestimate requirements in practice, either because it does not take into account shear heating that occurs during flow through runners and gates, or because 'setting up' is not instantaneous at the moment the demoulding temperature has been reached.

REGRIND RATES

For some injection moulding and extrusion applications and for some materials it is essential that virgin polymer be used (i.e. no reground stock should be present). However, for reasons of economy it is sometimes possible to incorporate some reground material without undue detriment to the product. What should, however, be borne in mind is that if a fraction of reground material is constantly being fed back into the hopper, some of this material may have been recycled many times. This calculation shows how this is related to the fraction of regrind used.

Theory

Let the fraction of regrind = f.

Hence
The fraction of compound reground

once	$= f(1-f)$
twice	$= f^2(1-f)$
n times	$= f^n(1-f)$

Also

The fraction of compound worked more than once =

1− (fraction of virgin material) − (fraction reground once)

And in general

The fraction worked more than n times

$$F_n{+} = f - (I{-}f)(f + f^2 + f^3 + \ldots + f^n).$$

Example

Let fraction of regrind $f = 0.5$ (a high value but not unknown where the sprues and runners comprise a high proportion of the moulding).

What fraction of polymer has been worked more than twice?

Substituting 0.5 into above equation

$$F_2{+} = 0.5 - (0.5)(0.5 + 0.25) = 0.125.$$

This means that one-eighth of the compound has been processed at least three times which would be excessive with some heat sensitive materials leading to discoloration, embrittlement or poor long term ageing behaviour.

If f is reduced to 0.1 then

$$F_2{+} = 0.1 - (0.9)(0.1 + 0.01) = 0.001.$$

STIFFNESS OF STRUCTURAL FOAMS

The use of blowing agents to produce structural foam mouldings provides a means of increasing stiffness, at the same time reducing the weight of the moulding, albeit at the expense of increasing volume.

This specimen calculation uses, as an example, the effect on the stiffness of a rectangular bar clamped at one end and loaded at the other as shown in the illustration.

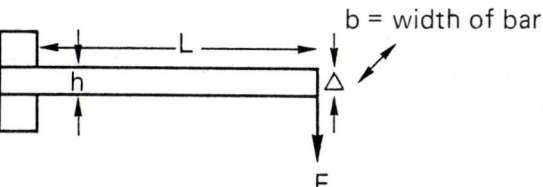

b = width of bar

Specifically, the calculation will show how much thicker the structural foam will have to be to maintain stiffness, and then determine the relative masses of the foamed and unfoamed parts. The method may be adapted for other types of loading using standard engineering calculations.

Theory The deflection Δ is given by

$$4FL^3/Ebh^3. \qquad (A.1)$$

Let us assume that the modulus of the structural foam is directly related to the proportion of solid polymer in the foam, i.e.

$$E_f/E = \rho_f/\rho = 1/e,$$

where ρ is the density, the subscript f refers to the foam and e is known as the blow up ratio.

Assuming that for the foam moulding F, L and b are unchanged, the deflection for the bar of structural foam will be

$$\Delta = 4FL^3/E_f \, bh_f{}^3 = 4eFl^3/Ebh_f{}^3 \qquad (A.2)$$

To calculate the thickness required to keep the deflection constant, i.e. $\Delta_r = \Delta$, equate (A.1) and (A.2) to give

$$h_f = e^{1/3}h. \qquad (A.3)$$

To calculate the mass of the foamed part as a fraction of the mass of the solid part of equal stiffness:

Volume of foamed part $= Lbe^{1/3}h.$
Hence mass of foamed part $= Lbe^{1/3}h\rho/e.$ (A.4)
Since mass of solid part $= Lbh\rho$ (A.5)

$$\frac{\text{Mass of foamed part}}{\text{Mass of solid part}} = e^{-2/3} \qquad (A.6)$$

Problem: Calculate the relative thickness and relative mass of rectangular bars with blow up ratios of 1, 1.2, 1.4, 1.6, 1.8 and 2.0.

Solution: It is simply necessary to substitute for e in equations (A.3) and (A.6).

Blow up ratios	1	1.2	1.4	1.6	1.8	2.0
Relative thickness	1	1.06	1.12	1.17	1.22	1.26
Relative mass	1	0.886	0.800	0.738	0.676	0.63

ADJUSTMENT OF DENSITY

It is frequently required to adjust the density of a polymer compound by addition of a second material such as a filler, a plasticizer or another polymer.

Theory

Let

density of original polymer compound = d_a

modifying material = d_b

required density = D

Now, density = mass/volume.

If mass of original material = M_a and mass of modifying material required to adjust density = M_b

Then

$D = (M_a + M_b)/[(M_a /d_a)-(M_b /d_b)]$.

Or, rearranging

$M_b + M_a (1-D/d_a)/[(D/d_b) -1]$.

Example

Let $M_a = 100$, $d_a = 1.0$, $d_b = 1.5$, $D = 1.1$. Hence by substitution

$M_b = 100(1-1.1)/[(1.1/1.5) -1] = 37.5$.

HEALTH AND SAFETY AND THE LAW

INTRODUCTION

In civilized societies there has been increasing recognition of the importance of health and safety, both at work and during leisure activities. In order to prevent irresponsible organizations and individuals evading their responsibilities in this respect, health and safety procedures have become increasingly backed by law. This section is intended as an introduction to the situation in the United Kingdom as understood by the author at the time of writing. For more detailed information the reader is referred to the actual legislation and also to the excellent manual produced by the Plastics Processing Industry Training Board, entitled *Standards of Performance: Health and Safety*.

STATUS OF HEALTH AND SAFETY PUBLICATIONS

The following is a summary of the legal status of publications on health and safety with particular reference to plastics processors.

Publication	Origin	Status
Statute, e.g. Health and Safety at Work Act	Made by Parliament	Full force of law
Regulations	Made by Minister of State	Full force of law
Codes of Practice	1. Approved Codes of Practice made or approved by HSC*	Not law but may be used as evidence in legal proceedings
	2. Industry Codes of Practice (produced by industry usually in liaison with HSE† and trade unions	Not law but represent best industry practice so set a standard for comparison

Publication	Origin	Status
Guidance Notes	Produced by HSE	Not law: guidance on acceptable practice
Health and Safety at Work Booklets	Produced by HSE	Not law: guidance on operation of laws and good practice
Other Standards e.g. BSI, ISO	Produced by various bodies	Not law: used as recommended standard by authorities in some cases

* Health and Safety Commission
† Health and Safety Executive

THE HEALTH AND SAFETY AT WORK ACT, 1974 (HASAWA)

The basic philosophy behind this act is that, 'Those who create the risks should themselves solve the problems that they create'.

HASAWA is an enabling act which imposes general duties on employers, and others, while still requiring them to comply with all current legislation which is relevant to their particular business. Such legislation includes the Factories Act, 1961, the Offices, Shops and Railway Premises Acts, 1963 and the various regulations making more specific requirements, for example where there are particular hazards involved in the use of certain types of machinery and substances.

The general purposes of the Act are to:

1 Maintain or improve standards of health, safety and welfare at work.
2 Protect the general public against risks to health and safety arising out of work activities.
3 Control the storage and use of dangerous substances.
4 Control certain noxious emissions into the air from certain premises.

RESPONSIBILITIES UNDER THE ACT

In addition to the employer, the Act also lays down responsibilities on employees, and de-

signers, manufacturers, importers or suppliers of articles or substances for use at work.

Responsibilities of employers

The prime responsibility for assessing risks and making arrangements to overcome them is placed on the employer. Under the Act they are required to undertake a number of actions which include the following:

1 Provision of a written statement of the company health and safety policy.
2 To provide a proper organization to implement the policy.
3 Plan, in writing, such necessary arrangements as reporting procedures and emergency drills and ensure that they are carried out effectively.
4 Install and maintain machinery, ancillary equipment and services such as electricity, water and gas in a condition which satisfies current requirements.
5 Devise work methods which do not expose employees to unnecessary risks.
6 Plan for the safe use, handling, storage and transport of all goods and materials and ensure good access and egress for all employees in all parts of the factory.
7 Provide every employee with adequate information, instruction, training and supervision, in relation to his/her job and level of responsibility.
8 Provide protective equipment and clothing.
9 Provide information about health and safety in the company to all employees.
10 Ensure that all training programmes include appropriate health and safety training.
11 Provide appropriate welfare facilities.
12 Ensure that in carrying out the work of the company there are no risks to the health and safety of the public (in particular neighbours and visitors).
13 Maintain the environment in a healthy condition tackling such issues as noise, temperature and ventilation.

Responsibility of employees

Under the Act certain responsibilities devolve upon the employees who are required to:

1 Take reasonable care that they do not endanger themselves or anyone else who may be affected by their work activities.

2 Cooperate with employers and others in meeting statutory requirements.
3 Not interfere with or misuse anything required to be provided in the interests of health, safety and welfare.

Responsibilities of designers, manufacturers, importers and suppliers

This group are required to:

1 Design articles which are intrinsically safe and without risks to health, or where this is not feasible design adequate safeguards for them.
2 Provide information on the safe use of such articles and substances.
3 Ensure that the article or substance is safe if used in accordance with this information.
4 Carry out (or arrange to have carried out) tests, if these are necessary to ensure that the article or substance is safe.
5 Carry out (or arrange to have carried out) research aimed to discover how risks to health and safety arising from the use of the article or substance might be eliminated or minimized.

Following the publication of the Act, a number of Regulations under the Act have been made. These include the following:

1 Safety Signs (1980)
2 Control of Lead at Work (1980)
3 Notification of New Substances (1982)
4 Classification, Packaging and Labelling of Dangerous Substances (1984)

THE FACTORIES ACT, 1961

The 1961 Factories Act consolidated the previous Acts of 1937, 1948 and 1959. It is now gradually being replaced by Regulations and Codes of Practice applying to all premises under HASAWA.

Amongst the provisions of this Act are a number specifying general working conditions, guarding of dangerous machinery, maintenance of hoists and lifts, boilers and air receivers, protection against dust and fumes and protection of the eyes.

Amongst Regulations issued under the Factories Act are the following:

1 Power presses (1965)
2 Asbestos (1969)
3 Highly Flammable Liquids and Liquified Petroleum Gases (1972)
4 Protection of Eyes (1974)

THE OFFICES, SHOPS AND RAILWAY PREMISES ACT, 1963

This Act was a particularly important piece of legislation concerning safety in offices, shops and railway premises as opposed to factories. The Health and Safety at Work Act has now made the 1963 Act of secondary importance but many of its provision still apply. As with the Factories Act there are provisions on such matters as general working conditions, guarding of dangerous machinery and safeguarding of hoists and lifts.

INDUSTRY CODES OF PRACTICE

A number of important Codes of Practice have been issued by the British Plastics Federation. Subjects include horizontal injection moulding machines, thermoforming and vacuum forming and the use of granulators.

APPENDIX 3

SOURCES OF FURTHER INFORMATION

Problems concerned with the design, manufacture and use of plastics components may be divided into three categories:

1 Problems essentially identical to problems already solved somewhere.
2 Problems whose solution is suggested by information already available.
3 Totally novel problems.

In practice most problems fall into the first two categories and in such circumstances recourse to various sources of information may be most rewarding. This section is divided into two parts:

1 Organizations within the United Kingdom concerned with plastics.

2 A selected list of textbooks and monographs on plastics.

ORGANIZATIONS

The Plastics Processing Industry Training Board, Coppice House, Halesfield 7, Telford Shropshire.

The PPITB provides training support for the plastics industry. It offers training courses on such subjects as injection moulding, extrusion and blow moulding at its own Training Centre at Telford, sets Standards of Training, provides advice on training to companies within scope of the Board and also provides a number of training grants.

The Plastics and Rubber Institute, 11 Hobart Place, London SW1W 0HL.

The PRI is the professional society for polymer scientists, technologists and engineers. It organizes numerous conferences, seminars and other meetings, issues a number of publications including a bi-monthly journal and awards qualifications in polymer science and technology.

The Rubber and Plastics Research Association, Shawbury, Shrewsbury, Shropshire.

RAPRA provides research, consultancy and information services (including abstracts) to member companies.

The British Plastics Federation, 5 Belgrave Square, London SW1.

The BPF is the organization for companies within the plastics industry. It represents the industry in discussions with government and other bodies and provides numerous services for its member companies.

BOOKS

The number of books concerned with plastics and related topics runs into thousands in the English language alone. The following list must be both highly selective and highly subjective. (Texts on design with plastics and chemical analysis are given in the chapters on these subjects in this handbook.)

GENERAL BOOKS

The most comprehensive source of information on plastics and polymers is the *Encyclopedia of Polymer Science and Technology* published by John Wiley, New York. The first edition, which was multi-volumed, was published in the 1960s. Volumes of the second edition have been appearing in the late 1980s and the edition will probably be complete before this book is published. In the view of this writer the new edition, although more up-to-date, is less comprehensive than the earlier edition.

The following provides a good introduction to polymers for graduate engineers:
Powell P.C. (1983) *Engineering with Polymers* London, New York: Chapman and Hall

PLASTICS MATERIALS

Brydson J. A. *Plastics Materials*, 1st edn 1966; 5th edn 1989. London: Butterworth.

An attempt by the author of this handbook to review the nature, history, preparation, properties, processing and applications of all types of plastics materials in one volume of about 850 pages.
Brydson J. A. (1988). *Rubbery Materials and Their Compounds*. London: Applied Science.

The equivalent for rubbery materials to the previous volume. Mentioned here since rubbery materials, and particularly the thermoplastic rubbers, are sometimes in competition with conventional plastics.
Elias H-G., Vohwinkel, F. (1986). *New Commercial Polymers-2*. New York; London: Gordon and Breach.

This is an excellent summary of developments in commercial polymers in the decade 1976–1985.
Elias H-G. (1977) *New Commercial Polymers 1969–1975*. New York; London; Paris: Gordon and Breach.

The predecessor to the previous book and a further useful source of information.
Margolis J. M. ed. (1985). *Engineering Plastics – Properties and Applications*. New York; Basel: Marcel Dekker.

Whilst typical of an edited book in that the specialist chapters are somewhat uneven, this volume provides a very useful source book on the major engineering thermoplastics.
Titow, W. V. (1982). *PVC Technology*. 4th ed. London: Elsevier Applied Science.

Although this 1233 page monograph is about PVC, there is a mass of information of use to processors of other materials. It is probably the best single volume on plastics processing operations.
Whelan A. (1982). *Injection Moulding Materials*. London: Elsevier Applied Science.

This contains a particularly useful chapter on properties and processing of the principal injection moulding materials as well as other interesting chapters on the effect of processing on properties and on quality control.
Whelan A., Goff J. (1987). *Moulding of Thermosetting Plastics*.

A small, privately printed, pocket sized softback of 72 pages, available from the authors or the Plastics and Rubber Institute. It contains a wealth of useful information on the compression, transfer and injection moulding of these materials which is not otherwise easy to come by.

Whelan A., Goff J. (1987). *Injection Moulding of Engineering Thermoplastics*.

Similar to the above but slightly larger (123 pages).

PROCESSING

There are few really comprehensive texts on polymer processing. Those that exist also range from highly theoretical and highly mathematical treatises, to very practical, but in some cases technically unsound, works. In the category of theoretical works the following may be mentioned.

Bernhardt E. C. ed. (1959). *Processing of Thermoplastic Materials*. New York: Reinhold.

Although over 30 years old and clearly very dated, this set a standard, which in the view of this writer has never been matched, with a clear attempt to apply theory to practice.

Brydson J. A. (1981). *Flow Properties of Polymer Melts*. 2nd edn. London: George Godwin (now part of Longmans).

An introduction to the subject of flow properties requiring A-level mathematics with some discussion of application to polymer processing.

Han C. D. (1976). *Rheology in Polymer Processing*. New York: Academic Press.

A more advanced text than the above, but with application to polymer processes.

McKelvey J. M. (1962). *Polymer Processing*, New York; London: Wiley.

Highly theoretical, but of high quality.

Tadmor Z., Gogos C. G. (1979). *Principles of Polymer Processing*. New York: Wiley.

Widely regarded as the best theoretical text on the subject, but requiring a graduate level of mathematics.

Tadmor Z., Klein I. (1970). *Engineering Principles of Plasticating Extrusion*. New York: Van Nostrand Reinhold.

Highly theoretical, but again of high quality.

Between the highly theoretical and the very practical books may be mentioned the Open University Course, PT614, Polymer Engineering. This course was introduced in 1985 at post-graduate level and may be taken as part of a Masters degree. The course includes eight text units and a number of video programmes. The written texts are an introduction to polymers, solid properties, polymer processing, extruded products, polymer composites, rubber products, injection moulding and advanced processing. There is considerable emphasis on the application of theory to solve technological problems.

At a much less theoretical level the Polymer Open Tech, which is associated with the PPITB, offers many units on aspects of polymer processing.

The Plastics Processing Industry Training Board also publish several very good loose leaf Standards of Performance Manuals designed primarily for shop floor training, but which have a wider interest. Several diagrams from these manuals have been used in this handbook. The manuals are being continually updated.

Monographs on individual processes include:

Elden R. A., Swan A. D. (1971). *Calendering of Plastics*. London: Iliffe.

Fisher E. G. (1971). *Blow Moulding of Plastics*. London: Iliffe.

Fisher E. G. (1976). *Extrusion of Plastics*, London: Iliffe.

Matthews G. A. R. (1982). *Polymer Mixing Technology*. London: Applied Science.

Rubin I. I. (1972). *Injection Moulding Theory and Practice*. New York: Wiley.

Schenkel G. (1966). *Plastics Extrusion Technology and Theory*. London: Iliffe.

Sweeney F. M. (1979). *Introduction to Reaction Injection Moulding*. New York: Technomic.

Whelan A. (1984). *Injection Moulding Machines*. London: Applied Science.

TESTING

In addition to the various standards quoted in the chapter on testing the following are recommended:

Ives C. G., Mead J. A., Riley M. M. (1971). *Handbook of Plastics Test Methods*. 1st ed. London: Iliffe.

Brown R. P. (1979). *Handbook of Plastics Test Methods*. 2nd edn. London: Applied Science.
Troitzsch J. (1983). *Plastics Flammability Handbook*. München: Hanser.

DICTIONARIES

Alger, M. S. M. (1989). *Polymer Science Dictionary*. London: Elsevier Applied Science.
While not for the novice, this recently published volume is bound to become a standard source of reference on more theoretical aspects of polymer science.

INDEX